Kids and Math

The Insider's Guide to Helping Your Child Make Sense of Math

Allison Shride

and

Suzie Shride

Copyright © 2017 Allison Shride and Suzie Shride

All rights reserved.

ISBN: 1539709485
ISBN-13:9781539709480

We are Allison and Suzie Shride and we want to help you get your kids back on track in math!

Imagine you could wave a magic wand and your child goes from hating math to being good at math, soaring his or her confidence and positively affecting all areas of their life and they tell you they're good at math and don't need help!

We can take you there!

I am Allison and I started Mathnasium of Parker 5 years ago. I was struggling to find meaning in my personal life and was dissatisfied with my successful career at that point. I was looking for something that satisfied my soul, something I could do by helping others. I happen to be blessed with a natural understanding of math and an ability to explain math in a way that makes sense to others.

One day while driving, I heard a radio spot for a Mathnasium franchise. I contacted the company, learned about the way they teach math and fell in love with it! Within 2 months, I had my center open and running! What I love about Mathnasium is seeing the impact that understanding math has on kids. I regularly have children tell me what a difference we have made in their confidence and grades, and parents share with me how feeling better about math has impacted many other areas of their child's life.

I am Allison's sister, Suzie, and I experienced exactly what you are experiencing with my own child. My daughter had been struggling with math for a couple of years and wouldn't let me help her, even though I was a classroom teacher at the time. In 6th grade, she came to me saying she hated math and just "didn't have a math brain." Allison's center was new, so I took my daughter to Mathnasium that summer. She enjoyed going and I thought we had found the golden key to math success but my daughter failed math her first semester of 7th grade. It turns out the problem was she not turning in her homework (which she did every night). At the end of that school year, my daughter asked to go back to Mathnasium over the summer. I thought, "absolutely! If my 8th grader wants to do math over the summer, I will make it happen!" Now, she's in an advanced math track in high school and a part-time math tutor. I saw what a difference Mathnasium made for my own daughter and I wanted to make the same difference for other kids, so I agreed to run Allison's second center in Littleton. It's a job I absolutely love! The Mathnasium Method is the best I have found for teaching math, and I get to experience daily the impact it makes on kids' attitudes, their success in school and their self-esteem. I wouldn't want to do anything else!

This book is for frustrated moms and dads who want to get their kids back on track in math. It's full of great tips we have found through working with hundreds of children and making math make sense to them.

Let's get started!

Table of Contents

CHAPTER 1 1

WHY DO CHILDREN STRUGGLE WITH MATH? 1

How Does Struggling with Math Affect Kids? 2

My Kid Skipped a Grade and Is Now Struggling with Math 3

What does Success in Math Look Like? 4

Why is Confidence in Math So Important? 4

What is Numerical Fluency, Why Is It Important and How Can I Help My Child Develop It? 4

What is Number Sense? 7

How Long Will It Take to Get My Kid Back on Track in Math? 12

My Child Has Good Math Grades, Can They Really Have Gaps? 13

Is Mathematical Understanding Really Necessary? 14

6 Signs Your Kid is Struggling with Math 19

CHAPTER 2 22

HOW DO I KNOW MY CHILD IS STRUGGLING WITH MATH AND WHAT CAN I DO TO HELP? 22

8 Tips for Helping Your Child with Math 23

Shouldn't Students Master The Basics Before Doing Problem Solving? 46

What If I Hate Math or Don't Understand Math? 49

How Do I Know What My Child's Learning Style Is? 52

Chapter 3 58

HOW CAN I HELP MY KIDS WITH MATH HOMEWORK? 58

So, How Do I Avoid Fighting Over Homework? 60

What Is the Biggest Mistake Parents Make When Helping Their Kids with Math Homework? 63

Why Do I Need to Know How to Support My Child with Math Homework? 64

  Here Are 6 Things You Can Do to Make Math Homework More Productive.. 65

  4 Warning Signs to Look for Which May Indicate Your Child Could Have Math Gaps 66

  Questions to Ask Your Child (Or Their Teacher) If You Think They're Falling Behind In Math Class 68

  My Tutor (Or The School) Rewards My Kid for Their Math. Shouldn't Kids Be Intrinsically Motivated? 69

  Are Online Websites Helpful and Can You Recommend Any? 70

CHAPTER 4 73
HOW DO I TALK TO MY KID ABOUT MATH? 73

  Here Are Five More Ways You Can Support Your Child's Confidence in Math.. 75

  What is a Growth Mindset? 82

CHAPTER 5 85
HOW DO I PLAY MATH GAMES WITH MY CHILD? 85

  What Benefits Do Math Games Have Over Other Instructional Strategies?.. 86

  How Often Should I Play Math Games with My Kid? 95

CHAPTER 6 97
HOW DO I FIND A MATH TUTOR?.. 97

  What Does Your Child Need?... 97

  What Do I Look For In A Math Tutor? 99

  How Often Should My Child Get Help? 100

  What Is the Difference Between Private Tutoring and A Math Learning Center?.. 101

  Other Things to Consider 102

  How Much Does a Math Tutor Cost? 103

What Does a Math Tutor Do? 103

How Will Hiring a Math Tutor Benefit my Child? 104

How is Mathnasium Different from typical Math Tutors? 104

CHAPTER 1

WHY DO CHILDREN STRUGGLE WITH MATH?

There are many reasons children struggle with math. In this chapter, we will discuss several of those factors including skipping a grade in math. We will look at how struggling in math affects children, what math success looks like, and why confidence in math is important. We'll also talk about numerical fluency and number sense and what those concepts really are. You'll get an idea of how long it could take to get your child back on track if they have fallen behind, why mathematical understanding is important and how your child may have math gaps even though they are getting good grades! Then we'll end with six signs that your child is struggling with math.

Have you ever watched a foreign film without subtitles? You can usually get a good idea of what is going on, but do you really understand the point of the movie? If a student is missing key foundational math skills, it makes it very difficult to keep up in class. They feel confused and frustrated...much the same way you feel while watching that foreign-language movie and not understanding the language.

Most schools treat kids the same because it is assumed that all kids learn the same way and at the same rate. The teachers are expected to cover vast amounts of material in a short period of time, not to mention CMAS or PARCC and other standardized tests. If your child doesn't master a skill for whatever reason – they missed some days due to illness, injury or travel, they were pushed ahead too quickly and skipped a grade, they had stressful life circumstances such as a death in the family or a divorce, etc. - well that's too bad. Teachers must move quickly to stay "on schedule". As a result, we see many kids fall behind in math.

What happens next? The class moves on! Here come multiplication facts, fractions, percentages, decimals, mixed numbers, exponents, ratios, x=what?, divisibility rules, algebra, geometry, equations. It never slows down so your child can catch up! They start to feel overwhelmed, embarrassed, and all alone! They begin to expect and accept failure as the "new normal" and think "I'm just not good at math!" They lose all confidence in themselves and all of their other grades begin to suffer.

Is it any wonder why you have to fight with them to do their homework? With many kids, it is easier for them to fight you and not do the work than to put effort into it only to fail again. After a while, it just doesn't seem worth the effort for them. They get overwhelmed and give up.

How Does Struggling with Math Affect Kids?

When a child starts to struggle with math, it becomes a downward spiral. They begin to feel stupid and ashamed. Embarrassment follows as they continue to struggle. They don't want to ask questions because they feel like they have a big "dummy" label on their foreheads that everyone can see and if they ask any questions, it will just prove they are stupid. Their confidence falls and it begins to affect all areas of their schooling and life.

My Kid Skipped a Grade and Is Now Struggling with Math

Pushing kids ahead, too soon is one of the unseen ways kids begin to struggle with math. Pushing students to the next level of math before they are ready is endemic in schools across the country, and is most pronounced in the move to have younger children take algebra. English and social studies teachers face the same problem when school officials, more interested in boasting about the numbers of kids in higher-level courses than in what they really learn, place students without the requisite skills in advanced placement classes.

Too much math, too soon is counterproductive.

The National Center for Education Statistics reports that from 1990 to 2007, the percentage of eighth-graders taking algebra went from 16% to 31%. California has been in the forefront of pushing kids into algebra: By 2009, 54% of its eighth-graders were taking algebra; the result of an initiative by California's State Board of Education. Why the early push? It's driven by the fact that some younger students wanted and were capable of more challenging math. But that's not true for all students.

In the lower grades, more time has to be devoted to practicing basic computational skills, so that they are internalized and eventually, come naturally.

Many of the concepts in algebra are abstract and if children are not developmentally ready to deal with abstraction, you can turn them off from math forever. Even the best students who can pull off A's in eighth-grade algebra by just memorizing eventually end up realizing they did not really learn it.

What does Success in Math Look Like?
Children who understand math are willing to ask questions in class, raise their hands and engage with the teacher and try new problems. They are willing to stand out in class. They understand how numbers fit together and can work with both whole numbers and parts of numbers (fractions, decimals, and percents). They have number sense and numerical fluency, and they are confident.

Why is Confidence in Math So Important?
It's not just important, it's critical! Kids need to see that they can do the work so that they can feel confident! This fresh, new confidence will spill over to many other areas of your child's life. At this point, you will start to see peace at home! You won't be having to fight with them to do homework. Not only will they have the confidence and the skills to attack their homework, but they'll start bringing home better grades from school. How do we know this? Because we have seen it over and over again with hundreds of students!

What is Numerical Fluency, Why Is It Important and How Can I Help My Child Develop It?
Numerical fluency is to math what literacy is to reading and writing. It's the ability to solve a math problem and explain how you got it and also, includes number sense and mental math. Numeracy does NOT involve school math, but merely the basic building blocks needed to handle numbers.

Parents have all heard how important it is to start reading with and to our kids at a very early age. Pediatricians recommend it. Pick up a book and read to your infant and toddlers as soon as, and as often as you can. This message has become a common knowledge.

So, why don't we treat math the same way? Why don't we, as a culture, start doing math with our toddlers? Our toddlers should be learning numbers and rudimentary math with the same urgency we have when we're teaching them the alphabet. And learning it long before they enter school.

Studies have shown a very strong, positive correlation between numerical fluency at a young age (preschool), and the likelihood of success later in life. Things like income potential, community involvement, even participation in the political process are strongly influenced by early age numeracy skills.

In 2007, a meta-analysis study of 35,000 preschoolers found that developing math skills early turns into a huge advantage for children. "The paramount importance of beginning school with a knowledge of numbers, number order, and other rudimentary math concepts, is one of the (findings) coming out of the study," said co-author and Northwestern University researcher Greg Duncan. "Mastery of early math skills predicts, not only future math achievement, it also predicts future reading achievement." The study goes on to say that the leading predictor of later achievement is school-entry math ability (followed by reading, and attention skills, in that order). Even in children with high levels of behavioral issues! According to the study, these skills are shown to have more influence on future success than social skills, or introverted versus extroverted personalities. In other words, a child's early age problem-solving skills are more important than how many friends they can make later on in life.

What, as parents, can we do to instill these very basic math skills in our kids? I'm not suggesting we start teaching toddlers the Pythagorean Theorem. No! Not necessary.

The "math" part of the brain acts much more like a muscle than a memory center. It must be worked, strengthened and stretched like a muscle. It loses proficiency when not used regularly. That's why school aged kids lose 2.6 months of scholastic math ability over summer breaks.

Start your kids with counting. Counting to ten (ten is a very important number to kids). Then, count items; Popsicle sticks, spoons, any common (safe) items. Have them count out ten to twelve items. Then, for example, take away three, now how many do we have? If you have seven spoons, how much more to make ten? And so on.

Use a deck of cards with the face cards removed. Flip a card; what number comes up? How much more to make ten? To make twelve? Flip another card, how much do you have if you double that number?

Count by different numbers. Count by fives and tens. Then, count by threes, fours, sevens, even twelves. Count by fives starting at the number three or starting at the number nine. As they start grasping these number concepts, stretch their math imagination, little by little. Introduce the concept of a half. Half of four, half of ten. Once they grasp that, introduce half of odd numbers. Half of one (a half is two EQUAL parts), half of five, etc.

A favorite game with small kids is the dice game. Explain to the child that the two opposite sides (of any standard six-sided) die always add up to the number seven. Knowing that, if I roll a die, and a five comes up, what's the (unseen) number on the bottom (i.e. two)? Now, roll two dice, and (without looking), what do the two bottom numbers add up to? Then, try three dice, etc. This requires not just adding numbers, but adding numbers that have to be imagined without looking. Mental Math! These little math games do not require a lot of time. One to two hours a week goes a long way.

Keep in mind, while it's ok to praise them for being smart, it's more productive to commend them for accomplishing and achieving these small achievements along the way to numerical fluency.

WHAT IS NUMBER SENSE?

Number sense is an intuitive sense and visualization of how numbers relate and how they can be manipulated. Like most skills, Number Sense can be developed through mastery of foundational concepts.

Among the most indispensable of these basic skills is counting & grouping or the ability to "see" numbers in groups.

Counting

Counting is the ability to recite numbers in order. It has been called the beginning of mathematics and is the next step after "knowing" numbers. As adults, we have been counting most of our lives and so it seems pretty easy. I mean, how difficult is it to go 1, 2, 3, ... etc.? Or even 3, 6, 9, 12, ...?

The purpose of counting is to assign a numeric value to a group of objects. The result of counting objects in a group does not depend on the manner in which the process of counting is conducted. For example, it does not matter how you count the number of eggs in a basket (one by one, by 2's or by 4's) or where you start counting (the left side then the right side or from the middle out), the result or the total number of eggs in the basket remains the same.

To say it another way, assume we are given two groups of objects. Take one object from the first group and move it to the second group. The total number of objects between the two groups remains unchanged. To obtain the total number of objects in the two groups, we agree to first count the objects in one group and then continue counting the objects in the second group. What if we start counting in the second group and then move on to the first? The total number of objects will still be the same.

Counting requires more than rote memorization. Many parents don't realize just how difficult counting is and how important it is for a child to master the most advanced counting concepts. Counting requires a child to understand the relationship from one number to the next. Counting skills develop in a predictable manner.

Young children start with the +1 counting relationship, meaning each number is one more than the previous number. The number the child ends on is the total. It requires the child to understand one to one correspondence. Common mistakes include assigning two objects one number, or not saying a number for each object.

Counting also helps children understand the concept of sequence, or in other words, that regardless of which number they use as a starting point, the counting system has a sequence or pattern.

Unlike kindergarteners, 1st graders must be able to start counting from any positive integer. That allows them to do addition and subtraction quicker and with fewer errors. 1st, 2nd, and 3rd graders learn skip-counting. Skip counting can be forwards and backwards by whole numbers. In other words, the relationship from one number to the next could be defined as +2 (2, 4, 6, 8 ...), or +8 (0, 8, 16, 24,...), or -2 (...8,6,4,2,0), or any whole number. Skip counting sets the stage for multiplication and division.

Children in grades 4 and up learn to count forwards and backwards by certain fractions (1/4, ½, ¾, 1...) and decimals (.25, .5, .75. 1). Negative number counting is introduced in 6th grade, requiring children to understand the concept of less than zero. This concept should be a prerequisite before getting a credit card or a cell phone!

Counting also incorporates the concept of quantity. The number represents the group of objects regardless of size or distribution. 9 blocks spread all over the table is the same as 9 blocks stacked on top of each other. Regardless of the placement of the objects or how they're counted (order irrelevance), there are still 9 objects. When developing this concept with young learners, it's important to begin with pointing to or touching each object as the number is being said. The child needs to understand that the last number represents that many objects and that the last number is also the symbol used to represent the number of objects. They also need to practice to count the objects from bottom to top or left to right to discover that order is irrelevant. Regardless of how the items are counted, the number in the set will remain constant.

Counting can also be abstract. This may raise an eyebrow but have you ever asked a child to count the number of times you've thought about getting a task done? Some things that can be counted aren't tangible. It's like counting dreams, thoughts or ideas, they can be counted but it's a mental process, not a physical process where you can touch and count.

When a child is counting a collection, the last item in the collection is the amount of the collection. For instance, if a child counts 1,2,3,4,5,6, 7 marbles, knowing that the last number represents the number of marbles in the collection is cardinality. When a child has to recount the marbles when prompted about how many marbles there are, the child doesn't yet have cardinality. To support this concept, children need to be encouraged to count sets of objects and then probed for how many there are in the set. The child needs to remember the last number represents the quantity of the set.

The ultimate goal of counting is to be able to count from any number, to any number, by any number.

To develop counting skills, parents can utilize the following exercises to help children learn to count from any number, to any number, by any number. These exercises should be done both forward and backward.

- Count by 1's, starting at 0 (0, 1, 2, 3…),
    - then starting at any number (e.g., 28, 29, 30, 31…).
- Count by 2's, starting at 0 (0, 2, 4, 6…),
    - then starting at 1 (1, 3, 5, 7…),
    - then starting at any number (e.g., 23, 25, 27, 29…).
- Count by 10's, starting at 0 (0, 10, 20, 30…),
    - then starting at 5 (5, 15, 25, 35…),
    - then starting at any number (e.g., 37, 47, 57, 67…).

- Count by ½'s, starting at 0 (0, ½, 1, 1½...),
  o then by ¼'s starting at 0 (0, ¼, ½, ¾...),
  o then by ¾'s starting at 0 (0, ¾, 1½, 2¼...).
- Count by 15's, starting at 0 (0, 15, 30, 45...).
- Count by 3's, 4's, 6's, 7's, 8's, 9's, 10's, 11's, 12's, 20's, 25's, 50's, 75's,
  100's, and 150's, starting at 0.

The benefits of this type of counting practice are strong addition skills and ultimately, the painless mastery of multiplication facts.

As counting skills begin to develop, fractions can be introduced. Long before introducing words like numerator and denominator, teach children that half means "2 equal parts," and have them use this knowledge to figure out things like:

- How much is half of 6? 10? 20? 26? 30? 50? 100? 248? 4,628?
- How much is half of 3? 11? 15? 21? 49? 99? 175? 999? 2,001?

As the ability to split numbers in half develops, add questions like:
- How do you know when you have half of something?
- Half of what number is 4? 25? 2 ½?
- How many half sandwiches can you make out of three whole sandwiches?
- How much is 2 plus 2 ½? How much is 3 ½ plus 4?
- How much is 7 take away 2 ½? How much is 7 ½ take away 2?
- How much is 2 ½, four times? Seven times? Two-and-a-half times?

- How much is a half plus a quarter?
- What part of 12 is 6? Is 4? Is 3? Is 1? Is 9? Is 8? Is 12? Is 24? Is 30?

These strategies can be started as early as kindergarten; however, they are appropriate for any person of any age who needs help with basic mathematics concepts and skills. The trick is to do these exercises both orally and visually, with little to no writing. As your child's Number Sense develops, the cycle of confusion, frustration, and intimidation that has developed in prior years can be broken, and now is a perfect time to get started!

Solid counting skills are needed to succeed in math. If your child is experiencing difficulty in math, it might stem from poor counting skills.

Calculators and computers have taken some the drudgery out of math, but they cannot replace the Number Sense that is learned in the process of mastering many of the fundamental concepts, facts, and skills of basic arithmetic. Problem-solving skills, the centerpiece of modern mathematics education, have their roots in this same development of Number Sense.

How Long Will It Take to Get My Kid Back on Track in Math?

This is a great question! None of us like to see our children suffer and struggle with something - especially when it affects their self-esteem and other areas of their lives. Unfortunately, how long it will take your child to get back on track depends on many factors including: If they have gaps in their mathematical foundation or not; how deep or wide those gaps are; what their current grade and/or math topic in school is; what your and their goals are for math; what their homework load looks like and what is their work ethic and motivation level to name a few.

As a general guide, we can tell you from experience that most students in the elementary grades who are 1-2 years behind in their math skills and who can work just on filling in their gaps at least twice per week will need 9-12 months of consistent effort to bring their understanding and mastery of math up to grade level. Middle school and high school students who are failing (or have failed) Algebra will need at least 4-6 months of DAILY work (6-12 months at twice per week) to bring their skills and understanding up enough to get B's or A's in their class (assuming they are turning in their homework and doing what is required of them in class). Students who work on their math inconsistently or who work very slowly or resist doing their homework will take even longer. Don't delay! The sooner you get your child help in math, the better it is for everyone!

My Child Has Good Math Grades, Can They Really Have Gaps?

Unfortunately, yes! Even students who have high marks in class, may not have a firm understanding of the concepts they are studying. They may be like Suzie, who got straight A's in math all the way through college calculus because she understood patterns and could memorize the algorithms and follow them. She really didn't understand what was behind the formulas until she started teaching math, and particularly, when she got trained in the Mathnasium Method.

So . . . Why should my child get math help, if they are getting good grades in math? Do they really need to understand what's going on? Yes (and no!) Because math builds upon prior knowledge and understanding, your child who gets good grades in elementary school may suddenly start failing in Algebra or higher math classes because of this lack of understanding. Or, their lack of understanding may result in a secret harboring of feelings of inadequacy you may not learn about until they decide to forego a STEM career for a less mathematically demanding one even though they have a natural aptitude for it.

Is Mathematical Understanding Really Necessary?

One of the lifelong benefits of mathematics education is the ability to solve problems. As adults, we are confronted with myriad problems daily and our capacity to look at, analyze, decompose or break the problem down, and come up with solutions derives directly from our understanding of math and our ability to think mathematically. "Most people think math is computation at the elementary level – drilling them in the skills," said Jeanie Behrend, an education professor focused on math education at California State University, Fresno. "Math is really about application and problem solving."

Note that I said our understanding of math. That is one of the differences between the Mathnasium Method and nearly all other math educational methods and curricula - understanding and mathematical thinking or number sense. The goal of math education should be for students to achieve understanding, and ideally mastery, of the concepts they study. Mastery is built upon three pillars: conceptual understanding, factual knowledge and procedural skill.

Students definitely gain factual knowledge and procedural skill when they practice math at school. Think times tables, Mad Minutes and other "drill and skill" practices. However, instruction needs to be directed beyond facts and procedures toward conceptual understanding; the lightbulb moment and the child's ability to think mathematically. This is best accomplished through Socratic questioning, gently guiding a child toward their own discovery of the answer, and by constantly having the child explain what they are doing and why. We say, "if you can't say it, you don't know it."

Many math teachers do not grasp the idea of conceptual understanding themselves. Far too many think that if students know the definitions and the rules, or algorithms, then they possess conceptual understanding. "It's easy for teachers to focus on memorization of facts and memorization of procedures without really identifying the important mathematics behind them and teaching those concepts to students," Behrend notes.

Memorization has never been the best way to learn math, but it was often enough to meet many of the old math standards. And knowledge of procedures is no guarantee of conceptual understanding.

"Doing" math is an operation. It's about applying mathematical procedures, or a sequence of steps and rules, such as addition, subtraction, multiplication, division, estimation, and measurement to solve an algorithmic or story problem correctly and successfully. It's all about the reproducing and applying facts and rules to achieve or attain that correct answer because, in the end, that's all that matters - getting the correct answer! This "mindless mimicry mathematics," as the National Research Council calls it, has come to be accepted as the norm in our schools. This still-dominant Old School model begins with the assumption that kids primarily need to learn "math facts": the ability to say "42" as soon as they hear the stimulus "6 x 7," and a familiarity with step-by-step procedures for all kinds of problems — carrying numbers while subtracting, subtracting while dividing, reducing fractions to the lowest common denominator, and so forth.

More than 70 years ago, a math educator named William Brownell observed that "intelligence plays no part" in this style of teaching math. Even now, most students are still being taught math as a routine skill. They do not develop higher order capacities for organizing and interpreting information. Thus, students may memorize the fact that 0.4 = 4/10, or successfully follow a recipe to solve for x, but the traditional approach leaves them clueless about the significance of what they're doing. Without any feel for the bigger picture, they tend to plug in numbers mechanically as they follow the technique they've learned. Drill does not develop meanings. Repetition does not lead to understanding.

As a result of the standard approach to math instruction, students often can't take the methods they've been taught and transfer them to problems even slightly different from those they're used to seeing. We see this all the time when we ask older students their multiplication facts. When we ask what is 12 x 12 we get the response "144!". Then we ask what is 12 x 13 and we all too often get blank stares or the comment, "I didn't learn what 12 x 13 was."

Another example is, a seven-year-old may be a whiz at adding numbers when they're stacked or arranged vertically on the page, but then throws up her hands in defeat when the same problem is written horizontally. She may possess a rich informal knowledge base derived from working with quantities in everyday situations that allows her to figure out how many cookies she would have if she started out with 16 and then received nine more – but regard that understanding as completely separate from the way you're "supposed" to add in school (where she may well get the wrong answer).

Math educators are constantly finding examples of how kids can do calculations without really knowing what they're doing. Children, given the problem (274+274+274) ÷ 3 set about laboriously adding and then dividing, missing the fact that they needn't have bothered — a fact that would be clear if they really understood what multiplication and division are all about.

Another example frequently observed is a question like this "A school bus holds 36 children. If 1,128 children are being bused to their school each day, how many buses are needed?" If you divide the first number into the second, you get 31 with a remainder of 12, meaning that 32 buses would be required to transport all the children. Most students do the division correctly, but fewer than one out of four got the question right. The most common answer is "31 remainder 12"

This sort of robotic calculation doesn't reflect a mental defect in the students but the triumph of the back-to-basics, drill-and-skill model of teaching math and that's not just one person's opinion. Analysts of National Assessment of Educational Progress data for the Educational Testing Service observed that students can "recite rules" but often don't "have any idea whether their answers are reasonable."

For example, many children learn a routine of "borrow and regroup" for multi-digit subtraction problems, however, they do not understand why that procedure works or how the "rule" came about. That would require a mathematical understanding of what we call the Law of Sameness as well as Place Value!

Another common conceptual problem is understanding that an equal sign (=) refers to equality — that is, mathematical equivalence or balance between the two sides of the equation. By some estimates, as few as 25 percent of American sixth-graders have a deep understanding of this concept. Students often think the equal sign signifies "put the answer here."

The idea of reading for understanding is clear enough (few adults, after all, spend their time underlining topic sentences or circling vowels), but how many of us have had any experience with math instruction that emphasizes understanding? We think of math as a subject where you churn out answers that are either right or wrong, and we may fear that anything other than the conventional "drill 'n skill" methods will leave our kids unable to produce the correct answers when time comes for them to take a standardized test. Indeed, it asks a lot for people to support, or even permit, a move from something they know to something quite unfamiliar.

If you want to grasp the poverty of your own education in math, I offer you the following challenge: explain long division. Explain it to a child, to an adult, to yourself – but really explain it. Use words to describe not just the process, but the reason for the process: why each number goes where it does; why you subtract, or divide, or bring down; why the process works. It won't be easy. I maintain that if you had been educated properly in math, it would be.

That is what mathematical understanding looks like. The ability to justify, in a way appropriate to the student's mathematical maturity, why a particular mathematical statement is true or where a mathematical rule comes from.

Our primary concern in teaching math should be not to confuse "doing" math with thinking mathematically. With the opportunity to use computers and calculators always at our fingertips, it is thinking mathematically that will give our children lifelong skills that will support them in every endeavor they attempt.

6 Signs Your Kid is Struggling with Math

The first step to solving a problem is identifying it. You've likely recently had a parent-teacher conference or you are about to meet with the teacher outside of conferences. How is your child doing in their math class? Do you have a nagging feeling that maybe they aren't understanding it as well as they could? Parents often tell us that their child is getting decent grades, but yet, they still seem to be missing some concepts.

Math builds from one concept to the next, so falling behind becomes worse as each year wears on. If your child is having a hard time keeping up in math, it's important to act now. Fortunately, there are many tell-tale signs that a child is falling behind—parents simply need to know what to look for.

Below are six things to look for that suggest a child is struggling with math.

1. Math grades are lagging, but the student has good grades in other subjects.

2. The child has low self-esteem about math, making comments like: "I'm just no good at math" or "I'm not smart in math."

3. Your kid is missing key milestones. Addition/subtraction should be mastered by end of 2nd or 3rd grade, multiplication and division by the end of 3rd–4th grade, fluency with fractions at the end of 5th–6th grade.

4. Physical signs like counting on fingers suggest poor retention of number facts and/or a lack of numerical fluency.

5. You get comments from the child's teacher about your child "not working up to his or her full potential."

6. Your child seems "bored" with math.

If a child shows any these signs, there are several things parents can do. Seeking help from the student's teacher is an obvious first choice. Offering to help the child more closely with their math homework is another. And, of course, enrolling in an after-school math program or hiring a math tutor is another option that brings professional help to the situation.

We've just looked at the factors that cause children to struggle with math including skipping a grade in math and not developing number sense and numerical fluency at a young age. We saw how struggling with math affects children including how it affects their confidence and why that matters. We looked at what math success is and what the concepts of numerical fluency and number sense actually mean. You've got a sense of how long it could take to get your child back on track if they have fallen behind in math and why mathematical understanding is important to your child and their future. You've learned that your child may have math gaps even though they are getting good grades and we gave you six signs your child is struggling with math.

In the next chapter, we will give you tips on how to help your child understand math.

CHAPTER 2

HOW DO I KNOW MY CHILD IS STRUGGLING WITH MATH AND WHAT CAN I DO TO HELP?

In the last chapter, we looked at some of the causes of math struggles and the effects falling behind in math can have on a child. Over the next several chapters, we're going to look at ways you can help your child with their math, even if you are not strong in math yourself, including 8 specific tips for helping your child now with math. The first four tips have chapters of their own while the last four are covered in this chapter along with a guide to roughly what your child should know at each grade and some resources for additional information. Let's get started!

Have you ever thought, "My child is struggling with math! What do I do? How can I help them?" I faced these same questions with my own daughter when she was in third grade. I knew she had a decent foundation in math, however, things weren't clicking with her teacher and she was starting to miss some concepts. To compound the issue, she was starting to resist my help. She's a bit stubborn like her mom! That's why I got started with Mathnasium!

Here are 8 Tips we've come up with to help parents help their kids with math.

8 Tips for Helping Your Child with Math

1. Help your kids with their math homework
2. Talk about math and use math vocabulary accurately
3. Play math games
4. Hire a tutor
5. Develop Mathematical Literacy
6. Give Math problems context and use actual items to manipulate
7. Try Different Approaches
8. Find Out How to Maximize Your Child's Potential

The first four tips are actually the last four chapters in this book. Go to the Table of Contents to find the page number of the tip you're looking for and read more!
Tips 5 to 8 are covered below.

Tip #5. Develop Mathematical Literacy. You read about math literacy and what it is in chapter 1. Here's an example. Emily is struggling with understanding her division problems, like knowing the difference between 12÷4 and 4/12. In third grade, she figured out that making 4 equal groups from a set of 12 is easier than making 12 equal groups out of a set of 4. When she got confused about the difference of 12÷4 and 4/12, she assumed she was supposed to make 4 groups from a set of 12. In fourth grade, that method didn't work so well because she was expected to be able to work with fractions. So sometimes, the problem really was 4 divided by 12, and she couldn't tell when.

Just like reading words, reading math requires understanding the grammar, context and word order. Division is especially difficult because it can look quite different from one problem to the next. Look at the difference in placement of the number "3" in the following sentences.

$$3\overline{)12}^{\,4} \qquad 12 \div 3 = 4 \qquad 12/3 = 4$$

They all express the same idea but with different symbols and without context, the meaning itself is ambiguous. They could mean four groups of three unidentified objects or three groups of four unidentified objects. The variety of symbols, various placements of numbers and lack of meaningful context in these problems are enough to make a child's head spin. If the child struggles with dyslexia or dyscalculia, the problem can be compounded even more.

Math literacy is one of the foundations of all math along with number sense, it is what all higher math concepts are built upon. See our chapter: How to Talk to Your Child about Math for more information.

Tip #6. Give math problems context and use actual items for your child to manipulate. What the heck is a manipulative? You've heard the teacher talk about using manipulatives in math class but you don't really know what she means. It is a fancy word for any object which is used or designed so a learner can perceive some mathematical concept by manipulating it - touching it, moving it around - hence its name. The use of manipulatives provides a way for children to learn concepts through developmentally appropriate hands-on experience, and they are very effective for children who are visual learners. Some examples that you can use at home include:

When working with money, use real coins and bills. Have your kids count back change, count money at the kitchen table and tell you how much they have, use money to make their own small purchases, play "store" at home using real money, etc.

For time, use a real clock with minute and hour hands. Turn off the digital clocks at home for a while. Ask your child for the time using the analog clock. Have them determine what time dinner will be served if it takes 45 minutes from now to prepare. Ask them what time they will get up in the morning, if they go to bed now (while looking at the clock) and how long they will have slept.

For fractions, use chocolate bars, graham crackers, pizza and other items that can easily be divided into equal sized pieces. You can also use small, equal-sized candies such as M&M's or Kisses and ask the child to share a certain amount evenly with you. Increase the number of parts (people) they are sharing the candies with once they understand sharing with two people. Have them share them with you and their sibling, or three of their best friends. Once they understand sharing in whole parts (8 cookies shared between 2 people is 4 cookies each or half of 8 is 4), have them share an odd number of pieces with you. For example, how can you and your child share 5 cookies between you so you each have the same amount?

Don't worry if they are creative in their sharing, i.e., they cut each cookie in half and then share out the halves. The beautiful thing about math is that there are many ways to get the right answer! Unfortunately, most elementary and some middle school math teachers don't grasp this concept because they themselves are not strong in math. Thus, one of the reasons teachers insist on children solving math problems only one way.

Use pictures. Even just having a child draw pictures to represent the problem can help the child make meaning of the problem. When working on a word problem, have your child draw what is happening. For example, if Sally has 8 flowers and Jane gives her 3 more, how many flowers does Sally have all together? Have your child draw 8 flowers and then 3 flowers. They might also draw Sally and Jane - that's OK! That's part of the fun for them - and keeping math fun is important, especially at a young age.

Repeat yourself as often as necessary. Kids are funny. Sometimes, a parent thinks they have explained a concept a hundred times and it is still not helping their child. Then, someone else explains it and all of a sudden, the child gets it. Why? Maybe a different phrase was used and it clicked with the child. Sometimes hearing "5 times 3" resonates differently than saying "3 groups of 5" or "5, 3 times." Different vocal tones resonate with children at different times. See Tip #7 for additional ideas. Or perhaps, the child just needed to hear the concept 101 times for it to sink in!

Think of repetition without expectation. That is being willing to repeat an explanation as many times as necessary until your child gets it without any demand that they know it by next week or winter break or the end of 3rd grade. It's hard but it works! When the pressure is off, kids feel comfortable enough to relax and learn the subject at hand.

Tip #7. Try different approaches or Ask someone else to explain it. Older children are often a great resource for helping their younger siblings understand concepts. They get it! They've just been through what their brother or sister is doing. Try pictures or manipulatives - see Tip #6. Try doing math mentally or written. Ask the question out loud. Have your child ask the question out loud and say their answer out loud. Sometimes, just hearing the problem causes something to click! Keep trying different approaches until you find the one that works. Be patient (see helping your child with homework).

Tip #8. Find out how to maximize your child's potential. Here are some tips.

If you're reading this book while your child is still young - congratulations! You've got time to make math fun and engaging for your child. When children are little, they are naturally full of curiosity and eager to learn new things. You can work with them on counting - by 1's, 2's, 5's and 10's. Helping them add and subtract items not just numbers is a great way to stimulate addition and subtraction fluency. For example: you can say "I have 3 apples, and I give one to you, how many apples do I have left?" Or "We have 2 dogs and our neighbors have 3 dogs, how many dogs are there altogether?" This helps children learn the language of math while having fun with easy, visual, and tangible objects. You can do activities like this anytime - while you're at home, driving in the car, or out on walks. Don't feel like you need to be sitting down with pencil and paper to do math!

If your child is older, don't worry - there is still plenty of time to get them engaged and learning!

The kitchen is a natural place to do math. Feel free to explore fractions and their relationship to each other with measuring cups and bowls. Have your child measure out 1/2 cup of something and see how many times they need to fill it up and pour it into a cup in order to fill up the cup. Next, do the same thing with 1/3 cup and 1/4 cup. How about tablespoons? How many of those are in 1/4 cup?

If you're cooking with a recipe, have your child help you figure out how much of each ingredient you would need to double or halve the recipe. This is real life math and young children love exploring and helping you figure it out! If someone in your household is a builder or seamstress, have your child learn to read a tape measure or yardstick and help measure either wood or fabric.

Play with money and coins. First, just give them your change jar (you have one, right?) and let them dump out the coins and count them. Suzie & I used to spend hours in our grandmother's basement counting out half-gallon ice cream cans full of dimes, nickels and a few quarters and half dollars. Have your child figure out 4 ways to make 10 cents, or 8 ways to make a dollar. Give them equivalency problems - how many nickels does it take to make a dime, how many dimes does it take to equal 6 quarters? When they are proficient with coins, start working on change back from a dollar, then $5. If they are doing well with this - play a game when you go to the store and try to figure out the correct change before the cashier can plug it into their cash register. This builds mental agility and a good understanding of how numbers work. Also, use the language of fractions here as well. Talk about a quarter being equal to 1/4 of a dollar. 3 quarters, or 75 cents being 3/4 of a dollar. How many nickels make a dollar? 20! Therefore, 1 nickel is 1/20 of a dollar.

Make sure your child is fluent in telling time on an analog clock. Again, this is a great place to bring in fractions. 15 minutes is a quarter or 1/4 of an hour. 30 minutes is half or 1/2 of an hour. Not only should your child be able to tell time, they should also be able to do elapsed time, i.e., if it's 12:00pm now, what time will it be in 2 and one-half hours? Alternatively, if you went to the store at 1:30pm, and got home at 5:00 pm, how long were you at the store?

Relate math to real life. If your child is a sports buff, talk about and figure out stats. How is a .278 batting average determined? What's your favorite team's on base percentage? What is a basketball team's winning percentage? What is your team's kicker's field goal average? If they are a shopper, figure out discounts on sale items. Determine if a $10 off coupon is better or worse than 10% off? Talk about how to figure out tipping percentages at a restaurant, or unit prices at the grocery store. If you're planning a trip, talk about miles per gallon and figure out how much gas you'll need to get there. Let older children help plan the budget for the family summer vacation.

As your child grows, talk about character development - the value of curiosity, determination and perseverance. Michael Jordan was kicked off his high school varsity basketball team and went on to become one of the greatest basketball players of all time because of his determination and work ethic. Albert Einstein was told at 9 years old that he would never be successful. Don't let your child give up or listen to others. Every child has the ability to be successful in math.

Be careful of pushing your child into accelerated math classes. This seems to be a big trend over the past several years and we've seen many kids who have been advanced or skipped through a class because of good test scores or teacher recommendations, only to struggle in the advanced class. We've worked with children who used to love math but have become frustrated, defeated or have started to dislike math because they were moved ahead when they weren't quite ready. Just because a child can do the math, doesn't mean they are mentally or emotionally ready for higher math classes.

Summer time is a great time for kids to forget a lot of what they learned during the school year. As parents, many of us are tempted to give our children the summer off from learning. This is a huge mistake as studies have shown that children lose as much as 2 1/2 - 3 months of knowledge during the summer. Play math games with your child, talk about math or enroll them in a summer math program to keep their skills strong or even advance them.

If your child is interested, involve them in STEM activities. There are lots of clubs now teaching coding for computers or apps as well as groups working on robotics, and engineering types courses for high school students. Many of the kids we see have a natural love for science. Foster that love and curiosity and encourage them to be solid in math as that is needed in many science courses.

Math is everywhere. Look for opportunities to do math problems beyond the paper. Engage your child in figuring our real life math and they will reap the benefits.

What Are the Basic Concepts My Child Needs to Know to Succeed in Math?

Unlike many other subjects your child learns in school, math builds upon itself. One concept leads to the next, and if a foundational concept is weak or non-existent, the structure is not in place for the child to succeed at higher math. These missing foundational skills and concepts are what are often referred to as gaps.

Dr. Wilson of John Hopkins University defines the 5 most important foundational concepts which are listed below, in the order in which they are typically learned.

1. Numbers
2. Wholes & Parts (Fractions and decimals)
3. Whole Number Operations (addition, subtraction, multiplication, and division)
4. Place Value
5. Problem Solving

Let's take a look at what each concept means and what it is to "learn" that concept.

Numbers

What does it mean to "learn" numbers? Understanding numbers is the ability to quantify objects. Learning about numbers is a preschooler's first step toward becoming a budding mathematician and in the early years, math learning is all about counting, number recognition and one-to-one correspondence. Look at this example: Two preschoolers are watching a parade. "Look! There are clowns!" yells Paul. "And three horses!" exclaims his friend, Nathan. Both friends are having a great experience but only Nathan is having a mathematical experience at the same time. Other children see, perhaps, a brown, a black, and a dappled horse. Nathan sees the same colors, but also sees a quantity - three horses. The difference is probably this: At pre-school and at home, Nathan's teachers and family notice and talk about numbers.

The more familiar one is with numbers and what they represent, the easier it is, generally, to see relationships involving numbers.

When Does Number Learning Begin?

When do children first become able to notice numbers? How important is it to notice and talk about them? Studies show that even infants are sensitive to quantity! Does that mean they "know" numbers? Probably only at an intuitive level.

Building on those intuitive beginnings are important. Every time you name a number, such as noticing; "Oh! I dropped three of the crayons," you sensitize children to numbers and teach a number word and its connection to a specific quantity. That is one-to-one correspondence. If you do it consistently, you are doing much more. You are encouraging children to think of the world in terms of, and to spontaneously recognize numbers. That is a gift that keeps on giving, because children can then create hundreds, or thousands, of mathematical experiences for themselves.

Explore Groupings

Parents and teachers need to be alert to naming small groups of objects and people whenever it is appropriate. "There are two airplanes." "Do you three friends want to play with the blocks?" Be especially alert to situations when naming small groups is important to the child. "You drew four baby horses! Are you going to draw four mommy horses?"

Of course, young children cannot yet recognize numbers in large groups. Unless they are arranged in certain ways, such as on a die, the limit is usually four to six. So, is recognition of numbers an early skill that fades away when real learning of numbers starts? The answer is no, for the following reasons:

1. Recognition of numbers supports the development of other number skills, such as counting. For example, one of the most important ideas about counting that many children do not develop is this: The last counting word tells how many. This, as mentioned in chapter 1, is called cardinality. Children will count three objects, but then, when asked how many, will re-count. But if they recognize groups of one, two, and three, then when they count out one, they see they have one, when they count out two, they see they have two, and when they finish and count three, they see three. They relate it to what they already know, and so the counting is more meaningful.

2. Recognition of numbers develops into more sophisticated abilities. The most obvious one is subitizing, or instantly seeing how many without having to count them. From a Latin word meaning suddenly, subitizing is the direct and immediate recognition of the number of a group. Simply stated, it's fast number recognition. It implies when someone shows you four fingers for only an instant, you recognize how many fingers they are holding up without counting. Many adults subitize the number of moves when playing a board game and no longer have to count each move. That fast recognition is important. For example, subitizing will later help children with adding. Many children add 4 + 3 by counting out four objects, then three objects, then counting all seven. The trouble is that their memory of the three and four on one hand, and the seven on the other, is too far away for the child to make a connection. Nevertheless, if a child subitizes the four, she is more likely to count on, starting with four, then five, six, and seven. Then, she learns a more sophisticated counting strategy and starts learning the fact that 4 + 3 = 7.

Seeing numbers in groups and being able to form meaningful groups from seemingly random assortments are skills essential for mathematical growth. So, be sure to notice and name numbers. Talk about how many objects appear in small groups everywhere around you. It's a sure way to put children on the path to math literacy, because it not only teaches them about numbers, but also ignites a mathematical way of thinking that will allow them to continue to teach themselves. And as you work with your child around numbers, teach them the concepts of denomination and quantity. The goal is to realize that everything in math (number, variable, term...) has a name and amount associated with it.

Wholes & Parts (Fractions & Decimals)

Once a child can count and understands the concept of one whole thing, we can begin to introduce the idea of wholes & parts. To understand fractions and decimals, a child must first understand the concept of wholes and parts. Knowledge of wholes and parts is the ability to see the wholes and parts in a given question and to utilize the ideas "the whole equals the sum of its parts," and "each part equals the whole minus all of the other parts."

Consider the number 6 as the whole number. It can also be broken down into various parts, 1 and 5, 1 and 2 and 2 and 1, 2 and 4, 2 and 2 and 2, 3 and 3. You may notice that these are essentially equations 1+2+2+1= 6 and 2 x 3 = 6. Trying to memorize all the equations that equal up to any given whole number would be ridiculous. The higher any number gets the more parts it has. A child who can break up any whole number into parts easily will find it easier do perform calculations such as adding, subtracting, multiplying, dividing and to use proportional thinking, work with fractions and many other math skills.

We always start with 1 something, be it 1 apple, 1 quarter, or 1 half-pound. That 1 initial something, no matter how big or small, is the whole. A whole always has at least 1 part: itself.

- If you have a 12-inch ruler, you could say that the whole is 1 foot, in which case an inch is a twelfth part of the whole. You could also say that an inch is the whole, in which case a foot becomes 12 wholes.
- If a quart is the whole, then a half-gallon is 2 wholes; but if a gallon is taken as the whole, then the same half-gallon

is only a part. If the half-gallon is taken as the whole, then "a half" is the whole. Fun, isn't it? Kids like this kind of illustrative wordplay, and it helps them painlessly retain the concept.

Remember, what is the whole and what are the parts depends on the perspective from which it is viewed.

There are many strategies to helping children experience the wholes and parts idea in math. Younger kids benefit from manipulating concrete tools such as unifix cubes (little plastic squares that fit together). Visual learners may remember wholes and parts concepts as pictures. Some common tools for visual learners include dominos. Dominos also help with subitizing. When children are ready for more abstract ideas with wholes and parts, fact families such as 10-2=8, 10-8=2, 8+2=10 and 2+8=10 are introduced. Dominos are a great tool for showing fact families. Legos can also be really helpful in teaching wholes and parts to second and third graders.

Wholes and parts in the younger grades are pretty easy to incorporate into everyday conversation. At dinner you might say, "We need 5 napkins in total. Choose some red ones and some white ones. How many of each will you choose?" As they get older, incorporating wholes and parts into the conversation might sound more like this, "Your birthday party guests include 3 kids who are vegetarian and 2 who are gluten free and 10 who eat all types of pizza. Please, order enough pizza of the right type to keep everyone happy."

Once a child grasps these ideas, we can introduce fractions. When a whole is divided (fractured) into a number of equal parts, we have what is called fractions. When a person encounters the word "fraction," the images "part of a whole" and "equal parts" must come to mind. The notion that a fraction results from the equal division of a whole must be second nature and deeply imbedded in the child's consciousness, and the earlier in life, the better.

A fraction is a part of a whole. The denomination (where the word denominator comes from) gives us the name of the fraction and is the number of equal parts in the whole. For example, if a whole is broken into 4 equal parts, the denomination (name) of the fraction is fourths. If we count 3 of those parts, then we have created the fraction three-fourths (3/4).

Even very young children can be introduced to the first fraction we start with, halves. If you give a young child a cookie and ask them to share it with you or their best friend, they will typically break the cookie into two pieces. The pieces may not be equal sized initially but you can have a conversation about making the pieces equal sized in order to be fair. Our grandmother used to help Allison & I understand this idea by telling us that the person breaking the cookie into pieces would not be the one picking the first piece! That was very motivating to make the pieces as close to even as possible!

After your child understands the concept of half, you can move onto other fractional parts. Start to include counting by common fractions (1/2s, 1/4s, 3/4s, 1/10s, 1 1/2s, 1 3/4s...), and later when they are ready, 4th-5th grade and older, by decimal fractions (0.1s, 0.5s, 0.25s, 0.01s, 0.75s, 1.5s, 1.75s...)

To understand what is going on in math, it is necessary to see the big picture of the subject at hand, the problem on which you're now working. Seeing this big picture nearly always involves determining the whole, its parts, and the relationship between them. By seeing how parts come together to form a whole and how the whole is broken down into meaningful parts, many aspects of mathematical problem solving can be learned very quickly.

For instance, when two parts form a whole, each part is called the complement of the other with respect to the whole. When three or more parts are considered, each part individually is the complement of the group of all the other parts.

Here's an example: In a certain class, there are 40 students of which 25 are girls and 15 are boys.
    The whole is 40 students.
    The parts are a group of 25 girls and a group of 15 boys.
    Then, 40 = 25 + 15 (the whole, 40, equals the sum of the parts, 25 and 15).
    The number of girls (25) and the number of boys (15) are complements with respect to the whole (40). Complements can also be thought of as "the rest of it—that which isn't there."

The two most common tasks in mathematics are:
1) Given the parts, find the whole.
"There are 25 boys and 15 girls. Find the total."
25 + 15 = 40

2) Given the whole and one or more part(s), find the other part.
"A box contains 12 marbles: 3 are red, 4 are white, and the rest are blue. How
many are blue?"

12 = 3 + 4 + (number of blue), so (number of blue) = 12 – 3 – 4

These two simple tasks, finding the whole and finding the missing part, are found in all branches of mathematics. The payoff for teaching children about parts and wholes early on comes later when we study:

Percent: one whole = 100%.
Probability: the probability of winning plus the probability of losing = 1.
Geometry: two complementary angles form a whole of 90 degrees. The Pythagorean Theorem or finding the area of an irregular shape also involve wholes and parts.
Calculus: the whole area under a curve equals the sum of the areas of all the thin
little rectangles, parts, that can be arranged to most completely cover that area.

Whole Number Operations

The whole numbers are the numbers which include zero and all counting numbers. The set of whole numbers can be written as {0, 1, 2, 3, 4, ...}. The whole numbers do not contain any decimal or fractional part. They consist of all positive numbers and zero. They lie on the right side of the real number line. Different types of operations can be applied on the whole numbers. The four basic operations with whole numbers are addition, subtraction, multiplication and division. In order to perform these whole number operations, a child must be familiar with numbers, be able to count and understand place value.

As your child becomes proficient at counting, teach them that number facts are just "the process of counting in groups."

Addition is "counting-up the total."
It is particularly important that children get sufficient practice to become adept with adding pairs of single digit numbers whose sums are not only as high as 10, but also as high as 18. For example, 4=4=8 5+5=10 9+9=18, etc. And it is particularly important that they get sufficient practice to become adept with subtracting single digit numbers that yield single digit answers, not only from minuends as high as 10, but from minuends between 10 and 18. Examples include, 18-9=9 12-8=4 10-7=3

Subtraction is "counting 'how far apart' two numbers are."
Children often do not get sufficient practice in single digit subtraction up to 18 to make it comfortable and automatic for them. Many "educational" math games involving simple addition and subtraction tend to give practice up to sums or minuends of 10 or 12, but not up to 18. Inadequacy of such practice and lack of "comfort" with regrouped subtractions tends to contribute toward a reluctance in children to properly regroup for subtraction because, when they get to the part where they have to subtract a combination of the above form, they think there must be something wrong because that is still not an "automatically" recognizable combination for them. Hence, they go to something else which they can subtract instead (e.g., by reversing the subtrahend and minuend digits in that column, so it will "come out" by allowing subtraction of a smaller digit from a bigger one) even though it ends up wrong. As an example, many children who don't have automaticity in subtraction, when tasked with this problem 24-6=? will reverse the 6 and 4 so it is 6-4 which they are comfortable with and gives them an answer of 2. Then, they bring down the 2 because 2 take away zero is

still 2 and they get an answer of 22 rather than 18. In a sense, doing what seems familiar to them "makes sense" to them.

Multiplication is "a fast way to count equal groups."
Since counting large numbers of things one at a time gets to be tedious, counting by groups of twos, threes, fives, tens, etc. is a skill to facilitate. Students have to be taught and rehearse to count this way, and generally they have to be told that it is a faster and easier way to count large quantities. Also, it serves as a prelude to multiplication, since counting by groups (of, say, threes) introduces one subconsciously to multiples of those groups (i.e., in this case, multiples of three), and, of course, grouping by 10's is a prelude to understanding those aspects of arithmetic based on 10's. Many teachers teach students to count by groups and to recognize quantities by the patterns a group can make (such as on numerical playing cards). This is important.

Division is "counting the number of this group that are inside of that group."
Our number system groups objects into 10s once 9 is reached. We use a base 10 system whereby a 1 will represent ten, one hundred, one thousand, etc. Of the counting principles, this one tends to cause the greatest amount of difficulty for children and leads to the next foundational concept, Place Value. Focus on 10, multiples of 10, and powers of 10 when working initially with your child on division.

Place value
Place value is one of the key concepts in mathematics curriculum and understanding place value (or not understanding it) will follow students through their mathematics journey.

Columnar representation of groups of numbers is simply one way of designating groups. We could designate groups with poker chips (see How to Play Math Games With Your Child), blocks, popsicle sticks, beads on an abacus and more. It is important to understand why groups need to be designated at all, and what is actually going on in assigning what has come to be known as "place-value" designation.

Groups make it easier to count large quantities; but apart from counting, it is only in writing numbers that group designations are important. If we are going to use ten numerals, as we do in our everyday base-ten arithmetic, and if we are going to start with 0 as the lowest single numeral, then when we get to the number "ten", we have to do something else, because we have used up all the representing symbols (i.e., the numerals) we have chosen -- 0, 1, 2, 3, 4, 5, 6, 7, 8, 9. We are stuck when it comes to writing the next number, which is "ten". To write a ten we need to do something else like make a different size numeral or a different color numeral or a different angled numeral, or something. On the abacus, you move all the beads on the one's row back and move forward a bead on the ten's row. What is chosen for written numbers in our system is to start a new column. And since the first number that needs that column in order to be written numerically is the number ten, we simply say "we will use this column to designate a ten" -- and so that you more easily recognize that it is a different column, we will include something to show where the old column is that has all the numbers from zero to nine; we will put a zero in the original column. And, to be economical, instead of using other different columns for different numbers of tens, we can just use this one column and different numerals in it to designate how many tens we are talking about, in writing any given number. Then it turns out that by changing out the numerals in the original column and the numerals in the "ten" column, we can make combinations of our ten numerals that represent each of the numbers from 0 to 99.

Now we are stuck again on a way to write one hundred. We add another column, and we can get by with that column until we pass nine hundred ninety-nine, etc. Thus, in our decimal, or base 10 number system, the value of a digit depends on its place or position, in the number. Each place has a value of 10 times the place value to its right. A number in standard form is separated into groups of three digits using commas.

The numbers in a written numeral have a name and a value, and both are essential for a student to understand the meaning of a number. As an example, consider these three numbers: 346, 463, and 634. These numbers all have the same digits but have very different values. The 6 in 346 represents six. Six ones. But the 6 in 463 represents sixty. Six tens. In the number 634, the 6 represents six hundred or 6 hundreds. In each case, it's six things — six units — but the unit you're counting changes depending on where the 6 is in the number.

Without this understanding, knowing when to exchange groups of ones for tens or how to handle a zero in the hundreds place when subtracting for example, confuses many students who then struggle with algorithms. They often struggle with knowing when to regroup ones and tens or "borrow," and the algorithms for adding and subtracting multi-digit numbers make little sense. However, being able to say that 635 is 600 + 30 + 5 does not, in and of itself, mean that a student understands place value. Place value encompasses not only position and value of digits, but also the composition of numbers and a number's relationship to others in the number system.

For students who struggle with understanding place value, zeros are especially difficult to comprehend. Take for instance the number 406. Many students read this as "four, oh, six". While this does not necessarily signify inadequacy in understanding place value, it does make it more difficult to understand what those three numbers mean when grouped in that order. Suppose the problem is 406 – 32; for a student that sees the middle digit just as zero, it becomes a puzzle as to how you could "take 3 away". On the other hand, a student who recognizes that 6 hundreds is the same as 60 tens, is able to make sense of taking three of those tens away.

The struggle with place value (and zero specifically) carries into multiplication and division. Many students try to use the standard algorithm for multiplication after being asked what 30 times 10 is. This is a good example of how strongly fluency and efficiency are related to understanding. If a student understands place value and base ten, they can quickly recognize that 10 groups of 30 is equivalent to 1 group of 300. Alternatively, students may know the "shortcut" of "adding a zero" without having an understanding of the underlying mathematics.

While these "shortcuts" may be helpful to students, they become problematic when students are faced with more difficult problems, or numbers involving decimals.

Decimals rely heavily on a solid understanding of base ten. Even a problem that seems to involve only whole numbers like 378 ÷ 30, requires students to understand where to record numerals in the quotient and how to record a decimal. In many cases, students will get 126 as their quotient (instead of 12.6) and not recognize that it is not a reasonable answer.

Just being able to use place-value to write numbers, perform calculations, and to describe the process is insufficient understanding. True understanding of place value is "the art of 'seeing' 10s." The goal is to be able to "see" how many 10s (100s, 1,000s...), 0.1s (0.01s, 0.001s...), and 1/10s (1/100s, 1/1,000s...) there are in any number (whole number, decimal fraction, and common fraction).

Problem Solving

The ability to solve problems is the gift that mathematics gives our children. Even children who are not interested in pursuing a STEM career or continuing their studies in a STEM field benefit from the study of mathematics because of their subsequent ability to solve problems.

What is problem solving? The dictionary describes problem solving as the process of finding solutions to difficult or complex issues. In mathematics, problem solving evolves from the basic - counting and grouping concepts (how many of these are there?) to the more complex problems.

The steps for solving problems are consistent throughout the levels of mathematics.

Here are the basic steps to solving any math problem:
1) Read and understand the problem.
    Re-read the problem several times, if necessary.
    Look for key words.
2) Plan how you are going to attack the problem.
    Determine whether you are looking for the WHOLE or for one of the
        PARTS.

3) Take into account the "limiting conditions" stated in the problem.

See if there are possible answers that can be eliminated due to

conditions stated in the problem.

4) Solve the problem.

Write and solve a mathematical sentence, usually an equation, that establishes the relationship between the WHOLE and its PARTS.

5) Answer the question.

Make sure your answer is reasonable.

Make sure you are answering the question being asked.

6) Check your answer.

Ask yourself the question, with your answer in place and see if it works.

As children advance in problem solving they should be able to:

a) Solve two and three step word problems using two or more operations.

b) Develop and use various techniques in Problem Solving:

1) break-down the problem into simpler parts,
2) apply the "easier number" method,
3) draw a picture,
4) make a table or chart,
5) work backwards,
6) use mental math,
7) use trial and error ("guess and check")

c) Check the answer if it is reasonable.

SHOULDN'T STUDENTS MASTER THE BASICS BEFORE DOING PROBLEM SOLVING?

Problem solving is a basic skill. In the past, too many students never got to problem solving because, they couldn't master a skill such as long division. Problem solving might just be the most valuable basic skill we can give to students. They'll certainly have problems to solve throughout their entire life!

There you have it! Mastering these concepts requires a systematic and logical sequence of instruction. Young children given the opportunity to master key math concepts, will be able to conquer the rigors of algebra and calculus later on in their academic career.

What Should My Child Know in Math at Each Grade Level?

Here is a rough idea of what children are expected to know and be able to do in math at various stages of their school career. The general idea behind the Common Core State Standards in Math is that there be a national set of standards so that all fourth graders across the nation have been taught or exposed to the same concepts. Thus, a child in fourth grade moving from Arkansas to California will have been taught the same material. The challenge to this idea of a national set of standards, is the fact that every school district and even schools within the same district have autonomy for selecting the curriculum they feel best teaches these standards. That autonomy results in a wide variance of what is taught and how it is taught. Therefore, your school may or may not follow the guidelines below, however, this list can serve as a reference to gauge whether or not your child is at a specific grade level. For specific information about the range of skills and concepts in school mathematics, please visit the Principles and Standards on School Mathematics on the National Council of Teachers of Mathematics' Website.

Pre-school - Kindergarten During this stage, children should begin to:
- count aloud
- compute the number of objects in a group
- understand that a particular number of objects has a fixed value despite the size or nature of those objects
- understand relative size and be able to sort objects by size and shape
- follow a sequence of two- and three-step commands
- be able to perform simple addition and subtraction computations

Grades One to Three During this stage, children should:
- begin to perform simple addition and subtraction computations efficiently
- master basic math facts (such as, 3 + 2 = 5)
- recognize and respond accurately to mathematical signs
- begin to grasp the concept of multiplication (grade three)
- understand the notion of measurement and be able to apply this understanding
- improve their concepts of time and money

Grades Four to Seven During this stage, children should:
- recall basic mathematical facts, including multiplication tables, with ease
- become competent with fractions, decimals, and percentages

- begin to understand the relationships among fractions, decimals, and percentages
- develop facility with word problems
- be adept at estimating quantities and rounding off numbers
- develop basic computer skills

Grades Eight to Twelve During this stage, children should be able to:
- employ an increasingly high level of abstract symbolic thinking
- perceive relationships and make translations among decimals, fractions, and percentages
- deal easily with a wide array of equations, formulae and proofs
- explain and illustrate mathematical concepts, rather than simply apply them
- plan and self-monitor during multi-step problem solving
- use calculators and computers with efficiently

In an attempt to foster their child's academic development, many well meaning parents may be tempted to provide flash cards as an initial learning tool. This is not the best method for teaching young children math skills with understanding. And it is pretty boring and tedious and may cause an enthusiastic, budding mathematician to get turned off on math.

With understanding comes the ability to connect the dots in math, and additionally, the ability to recreate the answer when memory fails! One of the best ways to help your child with math is to play games with them. See Chapter 6 for suggestions on different games you can play.

What If I Hate Math or Don't Understand Math?

If numbers are not your strong suit, how do you begin to help your child with their math homework without having flashbacks to the dreaded timed tests you took as a child? As parents, your children look to you for guidance when they are confused and struggling with their math. You don't want to appear inept in front of your children. More importantly, you don't want to confuse your children even more.

Many parents did not excel in basic math when they were in school. Others have not thought about higher concepts like algebra and trigonometry in years. Feeling inadequate about your own mathematical abilities does not need to stifle your attempts to help your son or daughter. These 5 approaches will help you assist your child with homework without causing you to break into a cold sweat.

1. Time, Empathy, and Encouragement

Take a few deep breaths and say, "It's been a while since I've done this math too. Give me a few minutes to look this over before I try to help you with it." Much of math anxiety comes from the added pressure of doing the math quickly. Taking the time to process the problem and analyze it deeply will reduce the stress.

Acknowledge that math can be difficult. Try to tackle it like something challenging you might do for fun, such as a puzzle. Simply letting your child know that you understand what they are feeling and can relate to their frustration is sometimes enough of a motivator to give your child that little push they need to keep trying. Remind them that although this challenge seems insurmountable now, with perseverance and patience, they will eventually master the task. Reminding them will also remind yourself to stay calm while working it out.

2. Review Prior Schoolwork

Math is typically taught sequentially. Give yourself the opportunity to look at prior lessons. Try reviewing previously completed worksheets or assignments that detail similar math problems. Walk through each step of the problem with your daughter or son, and have them explain to you what they think is occurring at each point in the example. Breaking the process into small chunks eases the anxiety. Take notes while talking through the process to help with comprehension. Big, complex, scary-looking problems don't look so overwhelming when you start with what you know and take it one step at a time.

3. Review Key Vocabulary

Unfamiliar words heighten anxiety. Don't know the difference between an equilateral triangle and an obtuse triangle? Google it! Almost any math word can be found online if you search with simple keyword phrases. Read the section Chapter 4 on Math Vocabulary.

4. Try Approaching the Problem Differently

Some learners are visual while others are auditory or kinesthetic. Kinesthetic learners learn best when moving and touching the material. Difficult math concepts don't seem as difficult when you approach them using your best learning modality. Try drawing a picture to see it or reading the problem out loud to hear it. Often, both children and parents respond better to alternative ways of thinking about the problem. When you think creatively about different approaches to the problem you not only lessen your own anxiety, you show your child alternative problem solving techniques.

Here are a few examples of different approaches.

- Basic fractions can be taught using pizza, fruit and other household objects. Using actual objects works very well with kinesthetic and visual learners.
- Word problems can often seem convoluted and misleading. Try representing the problem as a set of pictures to form a more simplified story to solve.
- Write out procedural steps for a multi-operational calculation before beginning to help your child reason through the process. Auditory learners may want to then read the steps out loud. Writing out the steps also aids in mathematical reasoning; a critical skill.
- Auditory learners will benefit from mnemonic, or memory, devices such as, "PEMDAS," to lessen the stress of trying to remember steps or information. "PEMDAS" stands for Parentheses, Exponents, Multiplication, Division, Addition, Subtraction and is used to recall the order of operations in math equations. Feel free to make up your own mnemonic device and get your child to help. Making it yourself helps make it more memorable. Make it silly. Nothing takes away stress like a good belly laugh. Here are a few common mnemonics to get you started:

- Steps of Long Division: Dad, Mom, Sister, Brother, Rover (Divide, Multiply, Subtract, Bring Down, Remainder).
- Order of Operations: Please Excuse My Dear Aunt Sally, (Parentheses, Exponents, Multiply, Divide, Add, Subtract), Left to Right (this latter is often left off and confuses children into thinking Multiplication always comes first).
- Letters and Values for Roman Numerals in Value Order: I Value Xylophones Like Cows Dig Milk (I=1, V=5, X=10, L= 50, C=100, D=500, M=1,000).
- Metric System Prefixes in Value Order: King Henry Died Drinking Chocolate Milk (Kilo=x 1000, Hecta=x 100, Deca=x 10, Deci=x 0.1, Centi=x 0.01, Milli=x 0.001).
- Trigonometry Formulas (where O=opposite, A=adjacent, and H=hypotenuse): Two Old Angels Skipped Over Heaven Carrying Ancient Harps (Tangent = O/A, Sine = O/H, Cosine = A/H). Or just SOH CAH TOA.

How Do I Know What My Child's Learning Style Is?

Think about the development of a few children you know. Did each child learn to walk the same way? Many children start with rolling, progress to crawling, and then to standing, and finally to taking those first wobbly independent steps at about 12 months old. Emmy took her own path to walking. She learned to crawl at 12 months old. She stood independently and took her first steps on the same day at 19 months old. Zoe skipped crawling altogether and started walking at 10 months old. Just as many children develop gross motor skills differently than their peers, many children develop mathematical concepts differently than their peers.

Are you concerned your child just isn't progressing in math at the expected rate? Many parents feel frustrated with the child, the school, and themselves. Unfortunately, the school system is not set up to cater math instruction to different types of learners. As children who learn math differently progress into higher grades, they can get farther and farther behind.

Your child likely sees math differently than you and may see it differently than the school is teaching it as well. That doesn't mean your child won't ever do well in math. It means he or she needs instruction at a different pace or in a manner that is tailored to his or her learning style. With the right support, your child can learn to visualize and mentally manipulate numbers in ways you never imagined.

The "Other" 40% Get Left Out

Have you ever tried to teach a group of people something? It could be anything - a dance, a computer skill, or a game. If you teach it too quickly, only a few people learn it. If you go too slowly, people get bored, distracted, lose interest and stop trying. Therefore, you try to find the perfect balance for the majority of the group. In the classroom, the pace and approach of a lesson is designed to meet the needs of about 60% of the students. That leaves about 10% bored and 30% confused. They are the ones who learn math differently.

"Why do children dread mathematics? Because of the wrong approach. Because it is looked at as a subject." Shakuntala Devi.

Shakuntala Devi, also known by the moniker "Human Computer," and other famous mathematicians see math as more than just a subject in school. They describe math in terms of poetry, music, logic, and even the language of the divinity. They experience and approach math very differently from the way it is taught in most schools. Many struggling children just need a different approach to unlock their mathematical brain.

Types of Math Learners

Research about learning generally agrees that every person learns through a combination of listening, seeing, moving, talking, reading, writing and touching. However, most people have a preferred method of learning. It is the way their brain processes the world and information best. That method is referred to as their individual learning style. Learning styles are classified as:

- Visual - Visual learners think in pictures and like math instruction with pictures and graphs. To solidify a new math concept in their brain, it helps them to write it down and draw pictures.
- Auditory - Auditory learners listen carefully, speak slowly and respond to music and rhythm. To solidify a new math concept in their brain, it helps them to explain it to others or create rhymes and mnemonic devices. "Please Excuse My Dear Aunt Sally" is an example of a mnemonic device to remember the order of operations of Parenthesis, Exponents, Multiplication, Division, Addition, and Subtraction (don't forget to include, LTR though - left to right).
- Tactile/Kinesthetic - Tactile learners should be using their hands and bodies whenever possible. Having them

physically sort objects or act out a story problem helps them solidify new math concepts.

Schools and teachers are improving math instruction to incorporate activities for all types of learners, but math instruction in our schools still heavily favors visual learners. Often, auditory and kinesthetic learners are trying to learn math in a way that doesn't suit their individual learning style. If the pace is either too slow or too fast, that compounds problems of boredom and frustration.

Mathnasium - A Solution for the 40%

At Mathnasium we recognize every child is different. Each child learns to apply the same mathematical concepts including:

- Counting
- Grouping
- Fractions
- Proportional Thinking (comparing amounts, sizes, etc...)
- Percent
- Algebra

Our instructional methods, however, are different from most classrooms. We don't have to worry about teaching to a group – we teach one-to-one in a group setting. We also do not have to follow a prescribed linear order of teaching one concept before another and we teach to five learning styles: verbal (auditory), visual, tactile (kinesthetic), written (combination of visual and kinesthetic) and mental. Nor are we on a specific time frame for having to teach certain topics.

For example, Julia has a solid concept of proportional thinking and can apply it to decimal places out to the hundredths fairly quickly in her head. It was easy for her to look at 9.65 and 9.05 and say "there is a six tenths difference." After she demonstrated that proficiency, we skipped over that section and allowed her to concentrate on grouping instead. Most classrooms don't have that flexibility. Julia struggled with division in school because it was presented visually and she needed to touch and sort the objects to understand what $40 \div 8 = 5$ felt like not what it looked like. After a few lessons using a kinesthetic approach to division, going at her pace, she discovered she could do division better than half the kids in her class.

Some very smart children do not learn math the way they are being taught in the classroom or they may be bored with the way math is being taught at school.

5. Enlist Outside Support

Getting outside support is a great way to stop a cycle of anxiety, stress, arguments, and further anxiety. If you and your child both dread math homework, and you both grow agitated and angry when you try to help, read our chapter on How to Find a Math Tutor or get in touch with Mathnasium!

In this chapter, we learned 3 of the 8 tips for helping your child understand math. We also looked at the 5 foundational concepts for kids to have a solid grasp on math - numbers & counting, wholes & parts, whole number operations, place value and problem solving. Finally, we got a general look at what children should know or learn in math at each grade level and gave you some tips on how to work with your child if you aren't good at or don't like math and how determine your child's math learning style.

If you're not quite ready to hire a tutor or outside assistance, or you feel fairly competent about helping your child with their math, you just need some tips on making math homework less stressful and more productive for everyone. That's exactly what our next chapter will discuss; how you can help your child with their math homework.

Chapter 3

HOW CAN I HELP MY KIDS WITH MATH HOMEWORK?

Helping your child with their math homework can be one of the most stressful home experiences. Often, math homework time (more so than with most other subjects) turns into a battle with arguments, tears and frustration all around. My own daughter has left the table in tears and gone to her room, slamming her door behind her. Why is this? In this chapter, we will look at why math homework is so challenging. We'll discuss how to avoid fighting over math homework and give you some strategies to implement to minimize the stress and battles over math! We'll look at what the biggest mistake parents make when helping their kids with math homework, how to make math homework time more productive, 4 warning signs indicating your child may have math gaps which you can observe during homework sessions, using rewards during math homework time, and why working with a real person is preferable to working online for struggling kids.

First, why is math so fraught with struggle at home? There are many reasons but the top two have to do with either the child's lack of confidence and skills level or with the parents' and sometimes it is both. Bear in mind, if your child is struggling in math (you may not even know they are), they already have a lot of pressure they're dealing with. Internally, they are feeling stupid and incapable. They are likely getting frustrated in class because they feel like they are drowning and don't know where to turn. They're afraid to get help and ask questions because then the teacher and some of their peers will know how stupid they are. So, they put on a brave face and muddle through it, hoping to make sense of it later at home.

Adding to the challenge, most of our children are way over-scheduled. In our attempts to give our kids the best, most well-rounded childhood, parents have them involved in so many activities they need a day planner to keep track of everything! There's dance lessons and soccer practice. They take piano, vocal or violin lessons and may be studying Mandarin, Japanese or Spanish on the weekends. Some kids are enrolled in two or three sports at a time, and most are year-round athletes going from football to basketball to baseball or field hockey, and softball! It's crazy and exhausting! They don't have time to relax and be kids, let alone take an extra 20-30 minutes to work on and come to grips with a math concept they didn't quite understand that day.

Of course, their parents have had a busy day too! Working and playing chauffeur to their children to get them to all the over-scheduled activities, they then have to put together a meal, plan for the next day, try to organize the house a bit and eek out a few minutes to help Johnny with his homework. Whew! It's exhausting listening to some families' schedules! One or both parent may also feel inadequate when it comes to math and every time Maria asks for help, it brings up insecurities they would rather not acknowledge. Or, they may have been good at fractions, algebra or trig in high school but it's been a long time since they've used those skills so they may have to spend precious time online trying to remember how to FOIL an algebra problem or add fractions with unlike denominators. That adds to the overall stress of the math homework scenario.

The opposite could be true. The parent may have a great grasp and understanding of math and they feel frustration because their child just doesn't get it! They can't understand how someone can't grasp a topic that comes easily and naturally to them!

Finally, add in the different curricula that's being used these days - from EveryDay Math to Singapore Math and Investigations including online text books and practice sites such as Khan Academy, Bridges or Agile Math, and many parents who are good at math, struggle to help their children the way the school wants it.

Put all of this together and it's no wonder math homework time becomes a battlefield!

So, How Do I Avoid Fighting Over Homework?

First of all, be patient. Remember, math builds on itself and it takes time to learn and master. This is especially true if there are any foundational gaps. Additionally, kids who are struggling with math very likely find the pace of the school curriculum to be too fast and so they already need extra time and the problem is being compounded. Having patience can be especially challenging for adults to whom math comes naturally or easily. They just don't understand why their son or daughter doesn't catch on as readily as they did.

Here's an idea of what today's math looks like to a kid who is struggling. Imagine you are at work and your boss just assigned you and your team a difficult task. You take a deep breath and give yourself a little pep talk. You remembered you accomplished something just as complicated last year. It took you a few late nights but with the help of your colleagues, your hard work, and perseverance, you ended up with a product of which you were proud.

Now your boss says, "You will be given one hour per day for a week to work on this task. If at the end of one week you do not have the task completed, you will get a poor evaluation. Then you will be given another equally difficult task that builds upon your success of this first task. We will continue on adding task upon task for the next 13 years."

You think, if you successfully complete each task, you will have no problem. But if you miss a few days of work or your concentration lapses a bit, you may not successfully complete a task. Or, what if one task just takes more than five hours to complete? Then every task after that will get more and more difficult because each task builds on the success of the previous ones. You remember playing the game Jenga™ and how a tower can get taller and taller but weaker and weaker. Eventually, the whole tower falls down. Now you don't feel motivated to start the first task to begin with. Would you stay at your job?

How is this Scenario Like School?

By the time a child enters 2nd grade, they have been in school for about 70 weeks. A typical math period lasts for one hour per day. In these 70 weeks, these 5, 6, and 7 year-olds must master the foundational math skills they will use for the rest of their school careers. If they do not get a solid understanding of counting forward and backward in kindergarten, addition and subtraction in first grade will be almost impossible. What happens when children struggling to add and subtract move on to second grade and get a lesson in adding and subtracting 3 digit numbers with "regrouping" or "carrying and borrowing?" How long until these precious children throw their chubby little hands up in despair?

Set Kids up for Success, not Failure

Difficult math concepts require hard work and time to understand. Complex, multi-step math problems require a child to concentrate and try several approaches. Giving kids time to really attack problems and try different approaches to concepts is critical to deep comprehension. When they fully understand one foundational concept, they are prepared to try more rigorous and complex problem solving skills. Then they advance to more abstract methods. Finally, they can start learning a new skill that builds on the previous skills.

Why doesn't the school handle this?

In a classroom with 25-30 children, each working at their own pace, some steps get condensed or skipped over entirely. When Colorado implemented the new State Math Standards in 2014, the range of topics that had to be covered in the course of a school year increased by about 25%! More topics, less time to go deeply into the topics or provide sufficient repetition for struggling students to master the concepts. The pace of math class now is quite fast compared to how parents experienced math at school and the length of time for math class has shortened.

Math curriculum probably progresses at an appropriate pace for about 12 of the 30 kids in a given classroom. Another 6 can manage but are starting to feel the pressure in math class. What happens to the 10 that need extra time and instruction to process each skill? They fall farther and farther behind the standard math curriculum. And the 2-3 kids who need less time start getting bored. The result is that many kids, even children who grasp math concepts quickly, end up with gaps in their understanding. We call this "Swiss cheese math understanding" because it is full of holes.

WHAT IS THE BIGGEST MISTAKE PARENTS MAKE WHEN HELPING THEIR KIDS WITH MATH HOMEWORK?

Probably, the biggest mistakes parents make when helping their kids with math homework is not allowing enough time for their child to process the problem before jumping in to give them the answer. We see it every day when we're assessing students. We'll ask what number is 10 more than 43? We observe the child start to think about the question and put together what they know in math. After 3 or 4 seconds, when they are just on the verge of giving us the answer, the parent - trying to be helpful and encouraging - blurts out "Come on Jenny! You know this!" disrupting the child's thinking process so they have to start again. Now 6-8 seconds have passed, and the parent, worried about not being a good parent, says, "Just think Jenny what's 43 plus 10?" and stares intently at their poor daughter. Jenny answers completely unsure of herself, "53?" and the parent sighs with relief and says, "good girl!" What's wrong with this scenario? Many things . . .

The adult did not let their child figure it out for themselves for one. Another is the question was not what is 43 plus 10? The question was designed to see if a child struggled with word problems, so by rewording it, the parent did not obtain the answer we were looking for. A third problem with this scenario is Jenny feels completely unable to think and figure problems out by herself, and so facing a test at school where mom or dad or their private tutor is not there with her, Jenny falls apart and fails the test.

When working with your child in math, we recommend asking the math question then waiting (and observing your child to see them thinking about it). Wait 20-30 seconds if necessary and just when you feel you can't stand the wait any longer, wait another 10-20 seconds. The beauty of letting a child figure out the answer themselves is they gain confidence in their ability to solve problems and the concept they are working on is more likely to stick because they engaged more areas of their brain as they struggled to come up with the answer.

WHY DO I NEED TO KNOW HOW TO SUPPORT MY CHILD WITH MATH HOMEWORK?

The main reason is because your children will very likely have math homework, and even if you have a tutor, there will be days when you will have to help out. The tutor may be sick. You may only be able to afford a tutor twice per week, not daily. You may have to let your tutor go or they may move and you end up in between tutors while you interview and hire a new one. There are many reasons why you will end up supporting your child with their math homework and we want to make the experience as stress-free and productive as possible for everyone involved!

Here Are 6 Things You Can Do to Make Math Homework More Productive.

1. Let your child move. Give your child a chance to move before starting their homework. Let them play with a friend (no video games), walk the dog, play ball, etc. The point is to get them outside in the fresh air for 15-30 minutes being active. They have been sitting at a desk for 6-8 hours already. Their brain and their body need a break!

2. Feed your child. A healthy snack after school keeps the blood flowing to young brains.

3. Create the right space. Have a designated spot in the house for homework. Make sure all the right tools are at hand (pencils, paper, calculator, math text books, tablet or laptop if needed, etc.). The space should be as quiet and free from distractions as possible (the kitchen table might not be ideal if mom or dad is preparing dinner). Soft, instrumental music can help children stay focused.

4. Set aside a time each day for homework. It is ideal if it is the same time every day because if it is the same time each day, it becomes a habit.

5. Practice patience. If you are helping them with their math, don't hurry their answers. Give them time to think. Wait until you're uncomfortable, then wait another 20-30 seconds. New concepts need time to develop. Rushing your child shortcuts this process. Don't answer your child's questions for them! Often kids are on the verge of discovering the answer themselves when we as parents rush in with the answer! This is a frequent cause of fights and tears during homework sessions.

6. Talk to your child's teacher. Go to back to school night, parent-teacher conferences or set up a separate appointment to know what's expected. Know how the teacher wants the

homework done. Many methods today are different from what you learned in school. Ask if they have a parent math night at school to support the parents in understanding the math curriculum.

These are just a few ways to support your child with their math homework and make it less stressful. Additionally, helping your child with their math homework can give you a good idea if they are starting to struggle or not.

The next time you help when your child does math homework, watch for these 4 warning signs that your child may have math gaps (and read 6 Signs Your Child is Struggling with Math in chapter 1).

4 Warning Signs to Look for Which May Indicate Your Child Could Have Math Gaps

1. Your child cannot explain how they got their answer or how they know it is correct (or incorrect). Explaining the steps in the process is not enough. Not being able to explain the process will significantly reduce your child's ability to apply the proper steps in complex word problems. He will also not see shortcuts or creative problem solving, because he will be locked into following a certain sequence to solve a problem. Children who can get right answers in the lower grades, but don't understand the underlying concepts, often experience difficulties in higher grades.

2. Your child doesn't notice when she gets incorrect or illogical answers. For example, if your child is asked to find 12.5% of 326 and they answer 407. A person with a solid understanding of percentages would immediately notice that answer simply

doesn't make sense. Everybody makes errors. If your child occasionally gets the wrong answer but continues, as if everything is fine, she may be following steps accurately most of the time, but not understanding the concept. Not understanding the concept makes finding and correcting errors very difficult. It also leads to huge issues as she progresses to more abstract math.

3. Your child uses his fingers to count or looks down or up at a white board in their head while doing math. Counting fingers or objects is perfectly acceptable through first grade because children still need a concrete example of numbers and the quantities they represent. By the beginning of second grade, children should be able to do adding and subtracting without their fingers. They should be starting to incorporate strategies beyond counting one-by-one. Some children become aware that finger counting is no longer acceptable and they count their fingers mentally. They often look down (or up) as they do this. If you see your child looking down or up and to the side while adding, subtracting, multiplying or dividing, ask him what he is visualizing. Counting concrete objects, even mentally, takes much longer and is less accurate than using other strategies.

4. Your child says "I hate math" or resists doing math homework. It's no secret that people like to do what they feel confident doing. If your child expresses frustration or dislike for math, it often indicates she doesn't feel confident in her abilities. She may notice that it takes her longer or more steps than her classmates to get the right answers. She may complain that the teacher doesn't help her or that math is just "dumb." Perhaps she is starting to feel overwhelmed

and doesn't want to admit it. She may be waiting for you to step in and help.

Questions to Ask Your Child (Or Their Teacher) If You Think They're Falling Behind In Math Class

If your child exhibits one or more warning signs they may be lagging behind or struggling with math, you need more information to diagnose the cause of the problem. Start by talking to your child. This list of questions will get you started.

Ask your child . . . .

- Do you use "manipulatives" in class and do you have access to them whenever you need them?

Manipulatives are the tools math teachers use to show a math concept at the concrete level. Common manipulatives include cubes, rods, beads, base 10 blocks, fraction strips, counters, coins and adjustable clocks. Even older children move from a concrete level to an abstract level in math.

Manipulatives facilitate growing their concrete understanding of complex math concepts. This aids their ability for higher-order thinking skills.

- Does your teacher go slow enough and repeat things often enough for you to understand?

Not every child processes language the same way. It's easy for a teacher to keep going ahead without realizing the quiet child in the back did not hear or understand.

- Does your teacher show you how to do the math?

Some children learn by watching. They need to get the instruction to match their individual learning style.

- Do you get time in class to practice new math skills?

Even the best math brains need time to wrestle with the skills independently. Like learning to swim, even the best explanation doesn't actually help much until you are in the water trying it.

- Does your teacher let you know right away if you get an answer correct when you are just learning something?

There is nothing more frustrating than getting a whole worksheet of math problems wrong because you practiced a new skill incorrectly.

- Do you ask questions in math class?

Some students don't feel comfortable speaking up in a group. They would rather remain confused than to speak up and ask a question. Also, if they are already feeling inadequate in math, they are unlikely to ask a question and risk feeling even more stupid.

- Do you feel smart in math?

Self-doubt affects how persistent a child will be when they face a challenge in math. Studies shows that persistence or grit is one of the biggest determining factors toward succeeding in any subject.

You will probably think of other questions to ask your kid, and some of these are perfectly appropriate to ask the teacher when you are at parent-teacher conferences.

MY TUTOR (OR THE SCHOOL) REWARDS MY KID FOR THEIR MATH. SHOULDN'T KIDS BE INTRINSICALLY MOTIVATED?

In an ideal situation, yes, all students (and people in general) would have sufficient self-motivation to get them through life's tough spots! However, we all know it is difficult to be motivated when you feel defeated or unsuccessful about something. Many children who struggle with math experience that feeling of defeat, shame and lack of success around math. That is why some of them start "hating" math. Mathnasium founder, Larry Martinek, said, "Children don't hate math, they hate feeling confused, frustrated, ashamed and the whole host of negative feelings associated with math."

We all like to get recognized for achieving goals. Young children literally shout out for positive reinforcement as they learn new skills on the playground. "Look what I can do Mom!" "Grandpa watch this!" The need for positive reinforcement doesn't go away. Just as in Facebook, adults have plenty of posts about their latest accomplishments so their friends and family members can shower them with accolades.

That is why we recommend congratulating, applauding and celebrating your child's successes in math. We do so on the soccer field or on the basketball court, "Great shot Johnny!" "Way to go Megan!" Why wouldn't you do the same thing with math (or any other academic subject?). We want our struggling children to start associating math with good, positive feelings. Sometimes, all it takes is a few small successes and little bit of external rewards and praise for a child's attitude about math to change and for their confidence in their own abilities to start to grow.

Are Online Websites Helpful and Can You Recommend Any?

While there are many online options available to help your child with math, nothing beats working in person with a math specialist. This is because working directly with a math

instructor can do many things a computer program cannot. Some of these include:
1. Knowing what your child's learning style is and being able to adjust to that.
2. With one-on-one interaction, the child can ask questions as they go and get an immediate response.
3. In person, the instructor can identify where your child is missing key components of a concept. For example, when learning order of operations, many children are taught the mnemonic device, Please Excuse My Dear Aunt Sally or PEMDAS, for remembering the order in which to solve an equation. What many kids forget (or are not taught) is that MD and AS are equal when there are not parentheses and need to be completed left to right. Online, it would just appear that the child doesn't understand the whole concept, not just the one piece of it. This can occur with many other concepts that have prerequisite knowledge that must be in place including borrowing and carrying. Many issues occur because children do not understand the concept of place value.
4. One-to-one instruction allows the tutor to build a rapport with the child. Once a relationship is established, the child feels more comfortable speaking up, asking questions and advocating for themselves.
5. A math specialist knows what the common assumptions or mistakes are in many concepts and they also know your child's unique pitfalls.
6. Having one-to-one interaction with a live person allows empathy to develop. As adults we can empathize with the many feelings children experience that go along with math struggles. These feelings may include frustration, anger, overwhelm, self-doubt, and self-pity. Good ways to show

empathy to the struggling child include: listening, asking follow-up questions, brainstorming solutions, spending quality time with them, and hugging them.
7. Working with a math instructor encourages the child to use mental math where appropriate.
8. A live person can adjust the pace of the curriculum and location to suite the child. Some concepts come more easily than others and that varies from child to child.
9. A math tutor can work with disabilities and learning differences such as ADD, dyscalculia, and aspergers among others.
10. A live person can reward and encourage kids appropriately.

Although working with a real, live person is typically the best solution for struggling children, there are many online math programs. Some of the better ones include: Khan Academy, IXL, and EngageNY. You may find others that work for your child.

In this chapter, we discussed the advantages to working one-on-one with a live person versus doing math online. We talked about the biggest mistake parents make when helping their child with math homework and gave some tips on how to avoid fighting over math homework and how to make homework time more productive. We also looked at why math homework is so challenging and gave 4 warning signs you can observe during math homework that may indicate your child has math gaps and whether to reward your child for math well done.

Next, let's look at how you can talk to your child about math because what you say and how you say it has a profound impact on your child's thinking, confidence and interest in math.

CHAPTER 4

HOW DO I TALK TO MY KID ABOUT MATH?

So far, we have looked at why kids struggle in math, how you can help your child understand math, how to find a math tutor and how to help your child with their math homework. In this chapter, we will review how you talk to your child about math. This is important for you to know because many parents unintentionally undermine their children's confidence in math just by the way they speak about math.

Parents frequently are uncomfortable with math or certain topics in math themselves and they often say "I am no good at math" or "I never understood Trig." We regularly hear, "I just don't have a math brain" or "I don't understand the way math is taught these days. I learned it differently."

Children hear those statements and take on the idea that it's ok not to be good at math or they think "who needs math anyway" and give up or don't try because "there's no point." Additionally, when kids are young, they want to emulate their parents so if their parent is not good at math and tells them so repeatedly, they don't want to be good at math either. They don't want to be better than their parent.

If there's one golden nugget of math advice we can pass along to help you set your family up for math success, it's this stop telling your kids you are bad at math! You are spreading math anxiety like a virus!

Here's some food for thought: most people see no problem admitting their distaste for math! In contrast, Petra Bonfert-Taylor, a professor of engineering at Dartmouth College and a 2016 Public Voices Fellow of the OpEd Project, a non-profit working to increase the range of public voices and ideas, stated in an online post "Few people would consider proudly announcing that they're bad at writing or reading. Collectively, we are perpetuating damaging myths by telling ourselves a few untruths; math is inherently hard, only geniuses understand it, we never liked math in the first place and nobody needs math anyway."

Now that you have the myths, here are some math facts:

- Math understanding is not exclusive to an elite fleet of geniuses—virtually every child (and grown up!) has what it takes to succeed at math.
- Mastering math is much like mastering other subjects: concepts click into place with consistent studying, effort, as well as solid instruction and guidance.
- There are many people with fond memories of math class . . . just as there are those with fond memories of art, English, history, and music class.
- Our entire world runs on math… from quirky instances of math patterns in nature to the math used to develop technology that supports our everyday lives. Knowing math makes life easier! Ask anyone who's ever had to calculate a restaurant tip or adjust a recipe on the fly.
- Math is the foundation for all STEM (science, technology, engineering, and mathematics) fields but is useful for all career paths! As an example, the clothes we buy

wouldn't fit as well or look as good if the designers didn't use math when making dress patterns.

As parents, you can help crush cultural math myths and set the stage for your child's success by promoting positive, realistic attitudes about math and making your household a math-friendly home! Include some math games into the rotation for family game night (Monopoly and Cribbage are two of our favorites!). Get your child involved in cooking and dive into concepts such as ratio and proportion. If you have a pet, ask questions like, "If Buffy eats 2 cups of food a day, and this 40-lb bag of dog food has about 130 cups of food, how many days will the bag of dog food last? How many weeks? How many months?" If you're feeling the math blues, we specialize in making math fun—reach out to us for ideas and a dose of inspiration!

If you struggled with math as a child, take heart—it is entirely possible to support a struggling child without tapping into your own negative experiences. Use the right language when speaking with your child: Math isn't "hard", but it does require practice. As you explore math with your child, do so with fresh eyes, an open mind, and a desire to learn. You'll find many opportunities to slay your personal math dragons, boost number sense, and have fun!

HERE ARE FIVE MORE WAYS YOU CAN SUPPORT YOUR CHILD'S CONFIDENCE IN MATH.

1. Be honest but open to finding solutions

You can say, "I wasn't good at math and I know math is important, so let's figure this out or let's find someone who can help you." Your child will sense that you empathize with their feelings of frustration and shame.

2. Talk about math whenever the opportunity arises.
Get kids comfortable speaking and listening to the language of math. Read earlier chapters about numeracy and number sense for examples of using math in every day language.

3. Use math vocabulary terms accurately when talking about math.
As parents and children work together to solve math problems, they need to discuss math concepts using math words or vocabulary. If either parent or child has a confusion about math words, <u>odds</u> are the discussion will <u>spiral</u> into an argument. Many math vocabulary words have multiple meanings. There is the everyday meaning and there is a precise mathematical meaning.

In an <u>average</u> household in the Denver <u>area</u>, parents help their children with math homework <u>multiple</u> times each week. Often, the homework session <u>reduces</u> the quality of the <u>relationship</u> between the parent and child. The frustration may be as a result of a <u>combination</u> of <u>factors</u>. A <u>fraction</u> of concerned parents will read our article, "Stop the Tears, Arguments & Whining During Math Homework" on the Mathnasium of Littleton and Mathnasium of Parker websites for help.

Math Vocabulary Words with Multiple Meanings
Using mathematical vocabulary with multiple meanings requires parents and children to engage in a complex linguistic skill known as "code switching." Code switching is using two languages within one conversation. Code switching happens in bilingual households when someone inserts a phrase from one language in the midst of a conversation happening in a different language. When it comes to math discussions, this happens even

in households where English is a first language. In this case, the two languages are the language of math versus the language of everyday English. The problem with code switching occurs when not everyone in the conversation is aware it is happening. This often happens when the child is unclear about mathematical meaning of a word.

How Words Differ in a Mathematical Context
Math words and vocabulary terms have precise definitions to describe numerical relationships. Sometimes the mathematical definition is similar to the everyday usage and sometimes the mathematical definition is quite different. Look at the underlined math words used in the first two paragraphs of this point. The words "multiple," "area," "average" "combinations," and "odds" have similar meanings in everyday usage as they do in math. This may trick the parent or the child into believing they are using the term in the same way. It is entirely possible one is using the word with a precise mathematical definition and the other is using the term colloquially. Often the parent or child does not even realize there is a precise mathematical definition. They are unaware the code switching is occurring.

Consider the following conversation as a prime example of this possible confusion, using the word "average".

Child: "I have to find the average price for soda. The prices are $1, $1.50, 1.50, $3 and $1.25, $2.00. My answer key says, $1.70. To me it looks like the answer is $1.50. Two sodas cost $1.50 and there are two cheaper and two more expensive. There isn't even one soda that costs 1.70!"

The child is thinking the word "average" means "typical" or "common." If the parent rushes in to explain how the kid did it all wrong without first explaining the mathematical definition of "average" – the sum of all numbers divided by the quantity of numbers - it will only serve to frustrate the child. Some math books use the term "arithmetic mean" instead of the word "average." "Arithmetic mean" would probably cause the parent to be confused too.

Generational Differences using Math Vocabulary
Math instruction is changing to reflect new standards. There have been shifts in curricula as a result of the state's adoption of the Common Core State Standards. As the instruction model changes, math vocabulary shifts to describe the concepts, tools, and methods currently being used. Often parents get confused when their children are using unfamiliar terms and vocabulary. Consider the following scenario showing confusion over changing vocabulary and methods.

Elliot tells his dad he has to add fractions for homework. He needs help and his dad remembers that to add fractions, the denominator has to be the same. Elliot says "Yeah, my teacher says I have to make a factor tree to find the least common multiple or LCM."

The father shrugs and asks, "What is a factor tree and least common multiple?"

The kid tries to explain it but he doesn't explain it well because he doesn't understand the concept well. The dad sees Elliot struggle and says, "When I was in school I used multiples to find the lowest common denominator and then added fractions. Let me show you my way."

The father and son struggle and both get frustrated with each other. After a while, they realize they are both describing similar ideas and methods and that both methods work fine. The dad walks away thinking "I don't get this newfangled math my son is learning." The son walks away thinking, "Why can't my dad just help me with my math without turning it into a big ordeal?"

Tips for Deciphering Math Vocabulary Words during Homework Time

- Ask your child to define the math words. The child can draw a picture or use words, but the definition should be mathematically precise and developmentally appropriate. Suppose you ask your second grader to define subtraction. Young children have the basic understanding that subtraction decreases the quantity. Your young child may define subtraction like "when you subtract, you are taking stuff away and the answer is smaller than the first number." You do not need to correct them with a lesson about subtracting negative numbers. Save that for a lesson a few years from now.

- Make sure you understand all the vocabulary yourself. Nothing is more frustrating to a child than having a parent who refuses to use the same terminology as their teachers and math books. If you are struggling to learn the current math vocabulary words, give us a call at Mathnasium. We stay up to date with the latest methods and vocabulary.

- Acknowledge code switching and multiple meanings. We use words with multiple meanings frequently without problem. As long as everyone understands the context, there is very little confusion. When adults ask for the area code for a phone number, they rarely get confused with the idea of a code indicating length times height (area). However, children and second language learners have a harder time with this. Be very explicit about the definition you are using when discussing math.

4. Empathize but don't excuse

Sometimes kids who are struggling in math get so overwhelmed that they stop giving the subject their best effort. As a parent or teacher who wants to see them succeed, you may feel empathetic to their struggle but get frustrated with their lack of effort.

Empathy means "The ability to understand and share the feelings of another." As adults we can empathize with the many feelings children experience that go along with math struggles. These feelings may include frustration, anger, overwhelm, self-doubt, and self-pity. Good ways to show empathy to the struggling child include: listening, asking follow-up questions, brainstorming solutions, spending quality time with them, and hugging them.

Excuse means "The release of someone of responsibility." Don't excuse your children from doing their math homework or getting good math grades at school. Never say, "Oh, that's okay. Some people just aren't good at math" or "I was never any good at math either." Those statements make the lack of skills seem permanent and unchangeable.

A child who gets excused from working hard in math will end up in a downward spiral. Lack of effort leads to lagging further behind. Lagging further behind will further erode the child's self-confidence.

Try saying, "I hear how frustrated you are. It must be so difficult to sit in math class and not understand what the teacher is saying. Let's work together to solve this problem." A child who gets both empathy and a push to work through the struggles will learn perseverance and get stronger math skills. When they eventually learn the material, they can look back on the experience to give them the confidence that they can conquer other challenges in math, and in life.

Working hard should be coupled with a strategy for success and leads to our fifth tip.

5. A Can-do Attitude is Key for Math Success

Children in math classes in Colorado have long been plagued with boring memorization of procedures and math facts. Procedures and math facts certainly help some people derive the answer to a simple arithmetic problem, but they aren't much fun. What's worse, many students (and some educators) are so bogged down in getting the correct answer to math practice, they forget that math can be a creative and joyous pursuit. Practicing arithmetic procedures should be like sports drills - helpful in isolating certain skills, but not a substitute for playing the game. The new Common Core State Standards in Math address some of the problems of focusing on procedures instead of reasoning. Until education fully embraces mindset, however, the standards don't do enough.

Mindset Matters

Carol Dweck, a research psychologist at Stanford University wrote a book about the importance of mindset called Mindset: The New Psychology of Success and it is revolutionizing the way people think about learning. A fixed mindset is when people believe ability or talent is unchangeable - someone is either good with numbers, or not good with numbers. A growth mindset is the belief that intelligence is influenced by exercising our brain. Brain and cognition scientists have proven that our brains, like our muscles, physically change with use. Scientists call this phenomenon brain plasticity. Unfortunately, some people have been slow to accept the mountains of research indicating brain plasticity or neuroplasticity. Dweck's research shows that people with a growth mindset, that is people who know intelligence, talent, and ability is flexible, outperform people with a fixed mindset in many arenas.

If parents want to give their children a gift, the best thing they can do is to teach their children to love challenges, be intrigued by mistakes, enjoy effort, and keep on learning. That way, their children don't have to be slaves of praise. They will have a lifelong way to build and repair their own confidence. - Carol Dweck

WHAT IS A GROWTH MINDSET?
Jo Boaler, a well-respected math educator explains in her book, Mathematical Mindsets: Unleashing Students' Potential Through Creative Math, Inspiring Messages, and Innovative Teaching, how historically math instruction in the United States has been especially plagued by a fixed mindset instead of a growth mindset, even by math instructors. Math instructors with a fixed mindset are more likely to let learning deficits fester. Instead of only valuing right answers, instructors with a growth mindset value effort, process, work, and even mistakes. When people

work harder, they get smarter - their brain grows. Just like when you workout harder at the gym, your muscles grow!

Imagine the following scenario in a fixed mindset classroom: Mason is in the fourth grade and has been struggling to learn long division. He keeps forgetting the process and getting the wrong answer. His teacher gives him extra practice division problems to do at home. Mason does them, but hates every minute of it. The day of the test comes and Mason gets a D on the math test. Mason slumps and thinks "I am no good at math" and his confidence takes a hit. The class moves on to learning fractions.

Now imagine this scenario in a growth mindset classroom: Brinley is also in the fourth grade and has been struggling to learn long division. She keeps forgetting the process and getting the wrong answer. Her teacher takes a different approach and gives her the challenge to come up with her own method of solving a division problem with multiple digits. Intrigued with the possibility, Brinley works diligently to find an alternative method. She uses creativity and mathematical reasoning. At the time of the math test, Brinley is still struggling. Her math test comes back with the note "let's keep trying new ways - you don't quite have it yet." Brinley does keep working and experimenting and finally creates a method similar that works for her. Her confidence soars.

How We Should Approach Math Instruction
Jo Boaler says math should include lots of creative challenges. When math students are allowed to think about math creatively, it not only increases their interest, it also increases their reasoning skills. Students involved in exploring and discussing math on a deep level grow their brain and increase their confidence. Boaler has a website,

http://www.youcubed.org   if you want to learn more about fostering a growth mindset in your child.

The Good News About Mindset

Don't fret if you, your child, or your child's teacher have not yet fully embraced a growth mindset. It is a process and the first step is to acknowledge where you are in the process. When math teachers are shown the dramatic results of children who are taught using a growth mindset, amazing things happen. Classrooms plagued by learning deficits have risen to the top of their state in math scores.

So there you have it! The way you talk to your child or around your child about math can have a profound impact on the way they approach math and even on the effort they are willing to put forth to do math. In this chapter, we debunked several math myths, gave you 5 tips on how to talk to your child about math including being honest but accurate, speaking about math often, using math vocabulary correctly, empathizing with your child and helping them develop a "can do" attitude. We also discussed what a "growth" mindset is and how to foster it.

Our next chapter will reveal how important it is to play math games with your child and give you some advice and instruction for different games you can play! Have fun!

CHAPTER 5

HOW DO I PLAY MATH GAMES WITH MY CHILD?

Playing games with your child has always been a great way to spend time together, and adding in games that involve math can be equally fun and rewarding and include the additional benefit of developing math skills along the way. I often tell parents that playing math games is similar to sneaking vegetables into a dish – it's good for children, and they often don't realize that they are learning because they are having so much fun along the way.

In this chapter, we will cover the benefits of playing math games and some of the many games we love! This is not an exhaustive list, there are many more games available that can support and enhance mathematical ability and thinking. Your kids might even have fun making up some games of their own!

There is a huge variety of games out there – you can play with a simple deck of cards or use a board game like Monopoly. There are games made specifically to address different areas of your child's math foundation and online games. You can strengthen your child's computational skills (addition, subtraction, multiplication and division) easily through a regular old deck of cards, a few dice or a box of dominos. You don't need anything fancy!

Many of the students we see in our center struggle with numerical fluency whether it is remembering their addition and subtraction facts or their multiplication facts. Numerical fluency is being able to quickly and efficiently add, subtract, multiply or divide. As we mentioned earlier, the challenge with numerical fluency is, even though a student may master it, if they do not continue to play with numbers and use number facts on a regular basis, they will quickly revert to counting one by one (on their fingers or in their head) and/or they will forget their facts. As an adult you have probably experienced the same thing yourself, especially if you use a calculator a lot. You may have forgotten what 25 + 7 is or what 9 times 8 equals.

We also said how using calculators in school for basic facts will also erode numerical fluency very quickly – we see it often with some of our advanced students – they have forgotten how to do long division, do computations with simple fractions, add or subtract integers and more because they are accustomed to using their calculators for these simple tasks. Just like your muscles lose strength and quickness when they are not used, the part of your brain used for quick calculations and numerical fluency loses efficiency and the ability to perform calculations that were once routine.

WHAT BENEFITS DO MATH GAMES HAVE OVER OTHER INSTRUCTIONAL STRATEGIES?

- Games are inherently motivating. Invite your child to play a game. You are more likely to get a "yes!" than inviting your child to work in a workbook or run through flash cards with you. It will probably even be a highlight in the day for both of you. Games offer a way to connect, talk, and laugh with your child. Keep the attitude light and fun.
- Math games using strategy encourage children to think about numbers creatively.
- Math games with an element of speed increase numerical fluency.
- Math games give parents an insight into a child's mathematical thinking.

- Playing the same game repeatedly gives the child the opportunity to keep improving and refining their speed, strategy, and skill. The same game can actually be more fun the more it is played as children see their skills increase.
- Children of varying levels can play the same game together with some rule alterations. The child with more advanced skills becomes a role model for the other one(s).
- Different games provide different opportunities for different children to shine.

We have found playing a daily math game for 5-10 minutes actually improves struggling children's fluency as fast as worksheets or workbooks!

Here are some games that we use at our centers and can recommend. We have included directions for the games with an asterisk (*) beside them after the list of game names.

*Multiplication War**

Math War is a fun variation on the classic card game that challenges players to perform a calculation (in this case, multiplication). The player with the highest product wins the hand. You can also play this as Addition War, Subtraction War, and even Proper Fraction War!

*Go Fish (for complements of 10)**

Go fish is a favorite game among younger children. Look for pairs in your hand that add up to 10, ask your partner for a card that you need, and if they don't have it – go fish!

Monopoly

A fun, family-friendly board game in which players move around the board buying and trading properties, building up properties by buying houses and hotels, and charging the other game players rent. Play the game with "dollar" bills only - the debit card version is not as educational.

Set
Set is a fun, portable (deck of cards) game that tests visual perception and pattern recognition by challenging players to choose three cards that all follow the same pattern or a different one.

Battleship
Battleship is an engaging game for older kids that tests memory and spatial reasoning skills. Based on a numbered and lettered grid, guess where your opponent's game pieces are positioned. First one to "sink" all of their opponent's battleships wins.

99 or Bust
This is one of our students' favorite math warm up games! It is one of the games where we have seen the most significant improvement in student numerical fluency and is always recommended as a way to keep facts fresh and help prevent the slide back to using fingers to add or subtract (we change the rules so that when the reverse direction card is played, we subtract the numbers rather than adding them).

Albert's Insomnia
Albert's Insomnia makes learning math fun. Albert can't sleep and he needs help counting his sheep. This game is a mental math game that requires players to think creatively and solve equations. Albert's Insomnia helps with mental math, math

facts, learning order of operations, critical and creative thinking skills practice and so much more. It can be played in small groups or collaboratively in a large group or classroom.

Zoom!
A math game with a forward/backward twist! Players add cards and watch out for wild cards that can send their scores zooming forward, backward or nowhere at all. Builds multiplication skills. Reinforces addition skills. Applies probability strategies. Aligns with learning standards.

7 Ate 9
It's as easy as 1, 2, 3. Players add, or subtract, 1, 2, or 3 to the number on the top card on the pile to determine if they have a card that can be played next. Sounds simple, but with everyone playing simultaneously, the options are constantly changing. The first player out of cards wins.

Blink!
Reinhards Staupe's Blink Card Game-The World's Fastest Game. It's faster than you think. Blink is the lightning-fast game where two players race to be the first to play all of their cards. Using sharp eyes and fast hands, players quickly try to match the shape, count, or color on the cards. The first player out of cards wins. Fast and portable, Blink is instant fun for everyone.

Spot it
This award-winning game of visual perception is for the whole family. Develops focus, visual perception skills, speech-language skills, and fine motor skills.

Swish
Designed to challenge spatial intelligence, Swish is easy to learn and addictively fun for players eight years and older. Get ready to stretch your mental noodle with this engaging card game.

Connect 4
This game is great for strategy. Challenge a friend to rule the grid in Connect 4 Classic Grid, the game where strategy drives the competition! Line 'em up and go for the win! Choose the gold discs or the red discs and drop them into the grid. When you get 4 discs in a row, you win. It's simple, fast, and fun. Master the grid.

Speed
"Speed" is a card game designed to teach children to multiply and how to multiply faster. Once familiar with the numbers in a game, multiplication becomes simple. The game consists of eight decks of cards (2's-9's) in which children learn skip counting. "Speed" is hands-on and multi-sensory, with short games to match children's attention spans. Through the repetition of the game, memorization happens naturally. The game is social and great for spending quality time with your child. "Speed" works great for all learning styles: Auditory (hearing), visual (seeing), and kinesthetic (feeling or experiencing with one's body). When playing "Speed" the auditory learner continuously skip-counts forwards and backwards either out loud or silently to her/himself. The visual learner sees the number cards laid out in a row and then sees them as they go up and down. The kinesthetic learner gets the body involved in the race trying to win the game.

24
This game provides fabulous math facts practice! It's so simple: Just make the number 24 from the four numbers on the card. You can add or subtract or multiply or divide using all four

numbers on the card but use each number only once! Each set contains 3 challenge levels and 48 cards. Builds strong mental math power!

Learning Wrap ups
Learning Wrap-ups offer a fun and unique way to help children memorize their basic math facts. Children wrap the string from the problem on the left to the answer on the right, and then turn the Wrap-up over to see if they got it right. Use a stopwatch to measure improvement and to see how fast they can really go. Choose from Addition, Subtraction, Multiplication, Division, and even Fractions. Each set contains 10 self-correcting keys and cover all the facts for that operation.

Shut-the-Box
This is a traditional game of counting, addition, and probability that dates back to the 18th century when the game was enjoyed by Norman fishermen after a long day at sea. Today the game is played with family, used in classrooms for teaching addition and probability.

Sumoku
Sumoku is a unique crossword-style game with numbers. Players add up their numbered tiles to a multiple of the number on the dice. Scores grow with every connected row and column. With 5 game variations and endless challenges, the fun quickly adds up with Sumoku!

Directions for card games:
These are the most common games we play at the Mathnasium Center.

Math War Games (Parent and Student)
- <u>Addition War</u>
  o Shuffle Deck and split in two; student picks half
  o Students and parent each turn over two cards

- o Student adds each pair and tells which sum is higher
- o Whoever has higher total wins the cards
- o If there is a tie, put those cards to the side; winner of next "hand" wins that hand plus the cards put to the side
- o When all cards have been used, player with larger stack of cards wins

Variations:
- o 3-Card Addition War
- o 4-Card Addition War
- o 4-Card Addition War, then student draws a 5th card and subtracts it from the total of the other 4 cards

- Multiplication War
  - o Same as Addition War, except student multiplies each pair of cards

Variations:
- o 3-Card Multiplication War
- o 2-Card Multiplication War with positive and negative numbers (black cards are positive, red cards are negative)

- Proper Fraction War
  - o Same as Addition War, except student makes proper fractions out of each pair of cards and determine which has the greater value.
  - o This one is more challenging to play at home.

Games for Two or more Children of Similar Skill Levels

- <u>Ten-Buddy Concentration</u>
  (Develops a crucial skill)
  o Before the game, take the face cards out of the deck.
  o Spread the cards on a table, face down.
  Depending on your preference, either arrange the cards neatly in rows and columns, or scatter the cards on the table in no particular pattern. Just be sure the cards are far enough apart that each card can be turned over without disturbing nearby cards.
  o The players take turns.
  - Each turn begins with a player turning two cards face up.
  - If the cards sum to 10, the player picks up the two cards, and takes another turn.
  - If the two cards do not sum to 10, the player turns the cards face down again, <u>without changing their position on the table</u>. This ends the turn.
  o The winner is the player with the most cards after all of the cards have been picked up.

It is a good idea to include one or both jokers in the game. Jokers do not match anything, and will remain face down on the table at the end of the game. If you don't use at least one joker, whoever wins the second-to-last set of cards will automatically win the last two cards.

For an easier, faster game
- Play with only three suits, plus the fourth 5. (There must always be an even number of 5's.)
- Play with only two suits.

- Play with only one suit plus an extra 5.

- Complements of 10 Go Fish
  - Remove the face cards from the deck, shuffle, deal each player 10 cards and spread the remaining cards out in between the players, face down.
  - Look for any pairs of numbers that add up to 10 in your hand (These are your complements) and lay them down on the table in front of you.
  - When it is your turn, ask the other player if they have what you need to make a complement. For example, if you have a 6, you would ask them if they have a 4. If they do not, they will tell you to "go fish." Draw a card from the cards face down in front of you. If you draw a card that makes a complement of 10, put the pair down in front of you and then it is the next player's turn.
  - The winner is whoever has the most complements laid down in front of them whenever someone runs out of cards.

- What's My Number? (Requires a referee)
  - Deal each child the same number of cards face down.

  Deal fewer cards for a shorter game, more cards for a longer game.
  - The children must not look at their cards.

  Have the children stack their cards, still face down.
  - At a signal, each child lifts a card from his stack, and without looking at it, holds it face out against

his or her forehead, so that others can see the card, but the child holding it cannot.
- A referee announces the sum of the two cards.
- The first child to call out the number of the card he or she is holding takes the "trick."
- The child who takes the most tricks wins. This game works well with a third child acting as the referee, instead of an adult. Try this game without keeping score. The kids enjoy treating each draw as a separate game.

Variation: Have the referee call out the product (rather than the sum) of the two numbers. (Remove the zeros from the deck before the game.)

Variation: Treat black cards as positive numbers and red cards as negative numbers.

Try both variations in combination. That is, have the referee call out the product of the numbers, treating black cards as positive and red cards as negative.

We hope this list has given you some ideas and inspiration as to how to have fun with your child while helping them master some math foundations.

We have not included any online games or apps as these are constantly changing and they tend to be more solitary type of games. There are plenty of great choices out there if you want to add some online games to your child's repertoire.

How Often Should I Play Math Games with My Kid?

You might be wondering how often you should play math games

with your child or children. We say everyday! Playing games creates wonderful family time and most of the games we mentioned take only 5 -10 minutes. It's a fun addition to your afternoon or evening activities. We've given you games for all different ages. Children as young as 4 or 5 can play many of them, while some are for children a bit older. We have students who are in middle school and high school that enjoy playing games like Blink, Zoom, 99 or Bust, 7 ate 9, and Speed at our centers. We believe you are never too young or too old to enjoy a good game!

So far in this book, you've learned a variety of strategies to help your child with math yourself. Let's say this isn't working or you don't have the time, or whatever, and you need outside help. That is what our final chapter covers, how to find a math tutor.

CHAPTER 6

HOW DO I FIND A MATH TUTOR?

We've discussed how to know if your child is struggling with math and how to help your child with their math. We've given you strategies for handling math homework at home and shared tips on how you talk about math affects your child. We've even shared some of our favorite math games with you, so you can support your child's math. Now, what if you aren't good at math yourself, or you just don't like math? Possibly your schedule is just too busy for you to dedicate the time and energy required to get your child back on track in math? Perhaps, you and your child are fighting over math homework or you're just looking for some extra support? Maybe your son or daughter needs some additional challenge in math because they are getting bored in class. These are just a few of the reasons you might look into hiring a math tutor. In this chapter we'll give you some advice on what to look for in a math tutor.

There are many options out there for you and your child. You can find many tutors and tutoring services online and in your local community. Some tutors will come to you, some you'll have to meet at the local library, and others will work in a learning center usually located in neighborhood shopping centers.

WHAT DOES YOUR CHILD NEED?

Are they at grade level, advanced or are you afraid they are behind in their math class? If they are at grade level and they are happy being there, perhaps they only need some additional support when that day's lesson doesn't make sense. This is a good time for them to go in and see their teacher before or after school for some additional help. They can also ask a peer or classmate for help. Another option is to seek out a tutor to help them with homework when it is challenging.

If your child is above grade level, they may be bored in their math class and need some additional or more challenging material to keep them engaged. Look for someone that can offer enriching material – going deeper into current subjects or advancing them further in their knowledge. Today's curricula are a mile wide and an inch deep, so there is plenty of room to expand a child's understanding and love of math without getting them too far ahead of the game.

Unfortunately, a great number of children are struggling and are lagging behind where they should be in their math. We've seen children that are behind a year or more in their skills and some who are as many as 5 or 6 years behind in their understanding of math concepts. For these children, math is a nightmare and they can easily feel overwhelmed. Getting help with tonight's homework may help them get a decent grade on their homework, but with holes in earlier skills, you can bet that they won't be able to retain the material long enough to do well on a test, let alone understand it long term. If your child is a year or more behind in math, it is critical that you find someone that is invested in helping them with their long term, mathematical development – someone who is able to go beyond tonight's homework. They will need someone who can identify exactly where the holes are in their foundational skills and create a plan to go in and fill them.

For example, we see many children who do not have mastery of their multiplication tables. You may think "so what – that's what calculators are for." How will they be able to do equivalent fractions without being able to quickly multiply simple numbers? Or, how can they find the Greatest Common Factor or Least Common Multiple if they don't know what 8x7 is?

Multiplication is the basis for so many concepts further down the line that it is critical that children know their basic facts. You'll want to find someone who can build this into their lessons if they do not know them – even if they are several years beyond when children typically learn their times facts.

WHAT DO I LOOK FOR IN A MATH TUTOR?

First of all, you want to find someone who is a math expert - someone who is involved in teaching math on a regular basis. If you have a high school-aged child, you'll want to make sure that the tutor or instructor can teach the higher level material as well as work on filling in earlier concept gaps. You probably don't want someone who tutors many different subjects. You know what they say – "Jack of all trades, master of none."

A question we get asked all the time is if a tutor should be a certified teacher. Our answer is no – there are many wonderful tutors who are not certified teachers (and some excellent tutors who are). The two key qualities to look for is their expertise in math and the ability to explain concepts in a way that makes sense to your child. Many brilliant mathematicians cannot step their knowledge down to make math concepts understandable to someone who is not getting it the way it is being taught.

Next, you want to ask if the tutor or the learning center has their own curriculum or do they just follow along with whatever your child brings home from school. You'll want to find someone with a time tested curriculum and teaching techniques that have proven results.

Another aspect to consider is if the tutor offers on-going assessment and feedback to both your child and yourself. This is important because as foundational gaps are filled in, kids' recall of other topics they learned but couldn't do because the foundation wasn't in place, can get triggered. Ongoing assessment helps identify these areas and allows that material to be removed from the plan. If a child already knows a concept, there is no need to teach it to them again. Additionally, regular communication with the parents as to how the child is doing is an important part of the process. How are your child's study habits? What type of learner are they and how does the tutor utilize their preferred learning style to supplement their lessons? Does the tutor notice indicators of a possible learning disability that should be looked into? There are many things that can be communicated to the parents beyond the specific math topic at hand.

One key factor to discuss with potential tutors is if they foster independence and how they do that. A challenge with one-on-one private tutoring (particularly in home) is the child develops what's called learned helplessness. You have probably noticed this yourself when working with your child. You may be sitting across from them while they work on their math, not saying a word when suddenly they start erasing their answer and their work. "What are you doing?" you ask. "My answer is wrong," they reply. And, yes, their answer is wrong but when you question them about how they know, you can tell they don't know. They might say, "Oh I just know," or "I can tell," but they cannot explain it to you. Somehow, through osmosis or the twitching eyebrow or a subtle sigh, you communicated to them that the answer was wrong. Another way to tell if a child has developed learned helplessness is when you ask them a question such as "what's 7 + 8 + 9?" and they answer, "23? no, 25? no 24?" making the answers into a question, waiting for you to nod or say "yes! That's correct."

There are other factors to consider when looking for a tutor but these will get you started on the right foot!

How Often Should My Child Get Help?

If your child is at or above grade level, once a week is probably sufficient, depending on what your goals are. Twice a week if your child wants to move ahead or has goals of going into a STEM career. If your child is struggling, they should definitely be getting help two to three times a week at a minimum. If less often than that, they may not get the consistency they need in order to be successful.

What Is the Difference Between Private Tutoring and A Math Learning Center?

Private tutoring is what most parents envision when they think about getting a tutor or help for their child. This is where one person sits with your child for their entire session and helps them problem by problem until their homework is complete. There are some inherent pitfalls with private tutoring, one of them being learned helplessness. Have you ever sat with your child during their homework and ended up doing too much of it for them? Telling them answers when they were stuck or correcting mistakes as they made them? It's human nature to want to jump in and fix mistakes as we see them happen, but it doesn't always serve our kids in the way that we intend it to. What ends up happening when we do this is our kids learn to look up at every step to see if what they just did is right. They develop the inability to move on and complete a problem or a page of homework without having someone looking over their shoulder and giving them feedback along the way. This is disastrous during a test when they must do it all on their own without having someone else to help them.

With a learning center, children will spend time one on one with a tutor or instructor getting help through directions and explanations. They then get the opportunity to work on problems independently and have their work checked once it's completed. This gives children the confidence to know they know what the next steps in the problem are and that they can complete it on their own. This translates very well during test time, because they have done many problems independently and successfully.

Learning center instructors are also there to help when a child does get stuck and does not know what the next step is. They are there to guide them until your child says "oh – now I know what to do" and can go onto complete the problem.

OTHER THINGS TO CONSIDER

Here are some other things to consider or look into while you're deciding who is the best fit for you child.

- Does the tutor or learning center guarantee results for your child?

- Do they give your child additional homework? Some parents want this. We feel that giving extra homework does benefit the child. We want to be sure that children are getting the right answers and just as importantly, that their thinking and processing is correct so that they will be able to continue getting accurate answers. How many times has a child had homework that they've done wrong – only reinforcing the wrong way to do things?

- Does the tutor and the curriculum develop critical thinking and problem solving skills? Many students learn or memorize a formula or algorithm and can plug numbers in to arrive at the answer, but really have no idea what the concept really means. Or, maybe the problem is presented in a format that they are not used to seeing. If they are not good problem solvers, this will trip them off. I know, as I used to be one of these kids.

- Do they teach for full understanding and mastery of concepts and offer ongoing assessments to allow your child to demonstrate mastery?

- Does the tutor or learning center create a tailored curriculum and learning plan for your child – one that addresses current issues as well as holes in earlier material?

- How does the tutor or learning center keep you updated? Do they send home written progress reports?

- Do you know who is working with your child? Have they been background checked? Most learning centers will do background checks on their instructors – it might be up to you to do one on a private tutor.

There are many things to consider when choosing someone to work with your child. You'll need to be clear on what your child's needs are and then find someone or someplace that can meet those needs. There are excellent private tutors out there, but going to a specialized learning center will take the guess work out of the process.

How Much Does a Math Tutor Cost?
Tutoring costs vary depending on the education and experience level of the tutor. Typically, college students charge about $30 an hour and teachers charge about $45 an hour. If you are looking for a calculus tutor expect to pay even more. A Google Search for math tutors shows a range of $25 - $105 per hour.

What Does a Math Tutor Do?
Typically, math tutors help with homework, assignments and help students study for tests. Tutors rarely do their own assessments to examine where a student lacks understanding and

they typically do not have their own curriculum to teach the topics that need shoring up.

How Will Hiring a Math Tutor Benefit my Child?

The combination of help on tests and homework may temporarily improve math grades. Children may ask math tutors questions they don't ask in class. This will help them clarify concepts they are learning in school at the moment. Often, however, the child's grades drop again once you stop paying for a tutor. This happens because the tutors do not teach concepts children missed from previous years. These gaps in understanding come back to cause problems later.

How is Mathnasium Different from typical Math Tutors?

Mathnasium assesses each child to find any gaps in understanding. Using the results from our diagnostic assessment, we provide individualized math instruction to actually improve math understanding. Since we have our own proprietary curriculum, we are not dependent on the schools for what to teach. This means we can help your child catch up, or get ahead, during summer break. We also help with homework and studying for tests to improve grades during the school year. Our instructors are gifted mathematicians who love working with children. They must pass very rigorous math tests in order to even be interviewed for a position. Then, they must be able to work with and make math understandable to a variety of students ranging from elementary kids working on basic numerical fluency all the way to high schoolers in trig or calculus. For one-to-one instruction, we are less expensive per hour than similarly qualified math tutors.

Hiring a tutor can be a daunting process. We hope this information gives you some direction and makes the process a bit less stressful. Having a tutor is great, however, there are going to be times when you will have to help your child with their math homework yourself.

Conclusion

Well, that's it! Now you have a lot of good information as well as tips and resources to help your child get back on track in math. You've learned why children struggle in math and how to know if your child is struggling. We've shared advice on how to help your children with their math homework, how to talk to them about math, how to play games that will help them with math and how to find a math tutor.

Remember, every child has the ability to be successful in math but our classrooms are not set up to help every child succeed in math. Teachers are doing their very best - Suzie knows because she was one - however, they are set up for failure in many ways. In any given classroom 40-60% of the kids need something extra to be successful in math. Often, it is simply a matter of time - more time and more repetition of the topic or skill. Our fast-paced curriculum does not allow sufficient time for those children to keep up.

Sometimes, a child has gaps in his or her foundational knowledge. They may have missed a day or two in 3rd grade when a critical topic was being covered, or perhaps a loved one died when they were in 4th grade and the stress and grief they felt caused them to be unable to focus and they missed important lessons. Perhaps a friend in first grade felt bad because your child was better than they were at math and so your child "dumbed" down out of compassion for their friend and missed skills. There are many, many individual reasons why a child may have math gaps but our educational system is not set up to allow teachers to diagnose math gaps and even if they could, they don't have the tools or the time to then remediate those skills.

On the other extreme, approximately 20% of the children in a given classroom, naturally understand math no matter how well or how poorly it is taught. Some of these kids are truly gifted in mathematical thinking and our classrooms are not set up to adequately challenge and stimulate these children. They often get "bored" of math and decide they don't like it because it is dull and monotonous. What a shame!

You may find your child still needs assistance, or perhaps you don't have the time or patience to work with them on your own. You've tried traditional tutors and found they didn't quite meet your needs. If this is the case, we would certainly recommend contacting a Mathnasium Math Learning Center near you and getting a diagnostic math assessment. This way, you will truly know what's holding your child back in math, and can then make an educated decision about what to do about it!

HAPPY MATHING!

Made in the USA
San Bernardino, CA
13 January 2017